Gold and International Finance

This book describes the history of gold as a financial instrument and discusses gold exchanges in the major markets. It also describes the history of the Chinese Gold & Silver Exchange Society (CGSE), its current organizational structure and membership registration system. The book also includes the development and growth of the gold market in Hong Kong and the role played by CGSE in the growth of the Chinese gold market. It includes a brief description of the CGSE in the twenty-first century – its current role and what it may play in the future.

The book explains factors that influence gold price and the mechanism of price formulation. It also describes the historic trends in the demand and supply of gold and the global inventory of gold, trends of the demand for investment holdings, jewelry manufacturing and industrial usage.

The book also compares the movements in gold price with inflation and analyzes the data on how gold provides a hedge against inflation. It also examines and explains the relationship between gold and the US dollar (USD) and the correlation between dollar index (value of dollar against 16 major currencies) and gold price. It explores in depth on the relationship between gold price, output and inventories and major economic indices.

This is a good reference for those interested in the comprehensive view of gold and its importance in the world economies.

Haywood Cheung has over 30 years of experience in precious metals trading, securities and futures brokerage and forex dealing in Hong Kong. He is currently the President of the Chinese Gold & Silver Exchange Society, the Chairman of Hong Kong Precious Metals Exchange Limited and the Permanent President of Hong Kong Precious Metals Traders Association Limited.

Routledge Advances in Risk Management
Edited by Kin Keung Lai and Shouyang Wang

Gold and International Finance

The Gold Market under the Internationalization of RMB in Hong Kong

Haywood Cheung

Routledge
Taylor & Francis Group

LONDON AND NEW YORK

First published 2017 by Routledge

2 Park Square, Milton Park, Abingdon, Oxfordshire OX14 4RN
52 Vanderbilt Avenue, New York, NY 10017

Routledge is an imprint of the Taylor & Francis Group, an informa business

First issued in paperback 2019

British Library Cataloguing in Publication Data
A catalogue record for this book is available from the British Library

Library of Congress Cataloging in Publication Data
Names: Cheung, Haywood, author.
Title: Gold and international finance : the gold market under the
    internationalization of RMB in Hong Kong / by Haywood Cheung.
Description: Abingdon, Oxon ; New York, NY : Routledge, 2017. |
    Series: Routledge advances in risk management ; 8 | Includes
    bibliographical references and index.
Identifiers: LCCN 2016035094 | ISBN 9781138807952 (hardback) |
    ISBN 9781315750828 (ebook)
Subjects: LCSH: Gold—China—Hong Kong. | Finance—China—
    Hong Kong. | Foreign exchange. | International finance.
Classification: LCC HG295.C62 H625 2017 | DDC 332/.0424—dc23
LC record available at https://lccn.loc.gov/2016035094

ISBN: 978-1-138-80795-2 (hbk)
ISBN: 978-0-367-35039-0 (pbk)

Typeset in Times New Roman
by Apex CoVantage, LLC

Contents

Illustrations

Figures

Tables

Preface

Historically neither the HKD nor HKD-related transactions have played a noticeable role in development of the gold market in Hong Kong. The people have become used to clinging onto the USD. It was only after the Sino-British negotiation for handover of Hong Kong to China that the HKD began to draw attention from the market, because it started experiencing drastic fluctuations in its value. The negotiations went on successfully. The HKD showed a strong trend of appreciation as Madam Margaret Thatcher was all smiles in the process of negotiations but it plunged from around HKD 5~6 to 9 per USD when the Iron Lady realized that China was different from Argentina and agreed to return the territory in 1997. At that time, huge volumes of HKD-denominated gold (Hong Kong gold) transactions were made. Today, the value of daily transactions is HKD 120~130 billion, compared with HKD 60–70 billion at that time (gold was then priced at USD 700–800/oz while it is now at 1,100–1,200).

Why is the Hong Kong gold market so active? A further investigation leads us to discover an important strategy ignored by many traders. Supply of HKD, then a regional currency, was limited by authorities; money supply was inadequate for larger volumes. When international investors turned their eyes on HKD fluctuations and aimed for more HKD transactions, they found it hard to operate. What should they do to speculate in HKD? International gold dealers and investors found gold a useful investment avenue at that time as HKD-denominated gold (99 Tael Gold from the Gold & Silver Exchange Society) could be used for arbitrage trade.

At that time, there was a strong negative correlation between Hong Kong gold and HKD. The fact that the HKD was in crisis and investors were looking for a hedging tool was the reason for depreciation of the HKD and appreciation of Hong Kong gold. Prolonged negotiations had frustrated investors and the HKD was weak but investors were able to acquire USD by estimating the price of gold in USDs and paying in HKD to buy the equivalent amount of Hong Kong gold. In effect this amounted to exchanging USD for HKD. By using this strategy, they could short sell large volumes of HKD. Large currency transactions were nearly impossible as authorities had imposed controls on the currency and the exchange rate (e.g. they could raise the overnight rate to make currency extremely intense, just like what they did when the 1997 Asian financial crisis occurred).

Conversely, when the HKD was in a strong phase (success in negotiations led to a strong HKD), investors started opting for Hong Kong gold as a way of acquiring HKD and paying in USDs to buy American gold to effectively swap USD for HKD. This implied a bullish position on HKD. The hedging remained unaffected by the risk of any change in gold price; investors were assuming only the exchange rate risk. After the closing of the gold position, there were either profits or losses. Big gold dealers who already had gold could expand profits by delivering the gold and completing the transaction.

The Hong Kong government announced the HKD-USD linkage on 15 October 1983. It was no longer possible to trade in currencies through the Hong Kong gold transactions and volumes shrank significantly.

Today when CNY is being internationalized, Hong Kong as a CNY offshore centre may play an important role. An active offshore market leads to increased widening of the margin between onshore and offshore markets. Diversified investment channels can activate the capital pool of CNY. The exchange society launched RMB Kilobar Gold in 2011, the sole offshore CNY-priced bulk commodity product in the world. But the capital pool at that time wasn't activated, with daily transactions amounting to as low as 10 billion CNY, mainly because the then onshore CNY exchange rate margin was limited to 1 percent. Profit from arbitrage trade was small and trading volumes were therefore restricted. As we keep calling for openness, now the margin has risen to 4 percent.

Historically the prosperity of Hong Kong has been linked to gold transactions while recessionary phases have always been linked to declines in gold trade and the consequent weakness of the HKD. This phenomenon came to a halt after the HKD was pegged to the USD. Now, the offshore CNY is just like the HKD which was freely traded in those years, and the arbitrage trades between CNY denominated gold and the currency should trigger significant interest in CNY gold transactions.

Free Trade Area (FTA) will also play a unique role during the process. The central government is expected to undertake a number of policy experiments in FTA, such as in Qianhai FTA. Foreign exchange controls have been relaxed as an important attempt to link Hong Kong, the offshore CNY market, in order to expand the offshore CNY market and by implication, expand CNY's role as the settlement currency in trading activities. This activated capital pool stands out as one of the most important stimuli for overseas investors to hold and invest in CNY. Bulk commodities are viewed as a good instrument and among these, precious metals have and offer noticeable advantages. The large gold warehouse we've established in the Qianhai bonded area has advantages of being located within the border and yet being outside the customs, which helps gold dealers and users avoid the need for customs clearance. The government has implemented total quota management and is able to stimulate currency circulation with physical gold.

In the meantime, driven by the B & R policy, this pattern will be continuously expanded until gold warehouses are established in B & R countries to meet market demand. CNY is used for gold transactions settlement to expedite the process of CNY being carried overseas.

Arbitrage trades in gold currency are on the way, across the four global gold markets, New York, London, Tokyo and Hong Kong. With such an excellent foundation and the continued push for the internationalization of CNY, Hong Kong will surely get the right opportunity in the near future to exponentially expand trading in gold.

This book is a celebration of my cordial relationship with Professor Kin Keung LAI as my supervisor and as a friend. I have benefited intellectually from his piercing insights and taste for good research. The perpetual encouragement and motivation he provided have enlightened me immensely.

I would also like to express my heartfelt gratitude, love and affection to my family. Without their incredible support, encouragement, perseverance, patience, contentment, uplifting spirit and loving kindness, I would not have completed the book so smoothly.

Haywood Cheung
Hong Kong

1 Introduction

1.1 Gold market in Hong Kong

Gold is the King of the Metal Kingdom and the idol of the world of economics; it plays an important role in global financial markets. The gold market in Hong Kong, now an international financial centre, has secured a key position in the international gold market. The Chinese Gold & Silver Exchange Society (CGSE), which has operated in Hong Kong for over 100 years, plays a vital role in Hong Kong's gold market. I was elected as President of the Exchange in 2010. As the exchange and the gold market have developed and expanded, we have keenly recognized the complexity and variability of the gold market. Research about the structure of the gold market and the factors that impact this market should help market participants avoid risk and promote stable and steady development of the exchange.

In addition, the CGSE has just started trading RMB Kilobar Gold, the world's first offshore RMB-denominated gold product, since 14 October 2011. The exchange is interested in further research about arbitrage in RMB gold market, which is essentially an interaction among a formulation of RMB exchange rate and how RMB gold products facilitate the Hong Kong RMB offshore financial market and accelerate the process of RMB internationalization. This is expected to help investors to explore the still-not-fully-open but yet attractive Mainland gold market.

1.2 The context of this book

Besides being a precious metal, because of its unique intrinsic value and stability, gold is treated as a special kind of currency, as well as a special kind of commodity, playing a unique role in financial and commodities markets. As an important financial asset, gold has a longer history than most other financial assets. For centuries it has been a medium of exchange, measure of value and way of storing wealth.

The global gold market is an important part of financial markets and because of its special properties and the stable intrinsic value, it is often used to hedge financial risks, which is different from other hedging instruments used in financial markets. Market price of gold receives global attention and is affected by many factors.

Gold markets are mainly located in Europe (mainly London and Zurich), Asia (Hong Kong, Shanghai) and North America (New York). The main trading centres are London and New York with Shanghai also becoming of increasing importance. Second tier might include Zurich, Hong Kong, Singapore, Tokyo and Sydney. Including the key physical gold hubs would definitely add Zurich, Hong Kong, Dubai and, possibly, Singapore to the list. In terms of gold mining equities trading, Toronto and to some extent Vancouver should be added. Trading hours of major gold markets are such that gold trading continues almost uninterrupted, moving from Hong Kong to London to New York through the day.

RMB Kilobar Gold launched by the CGSE is the world's first offshore RMB-denominated spot gold contract. Institutional and retail investors may engage trade in RMB Kilobar Gold by opening an account with CGSE's AA grade members. To effectively mitigate the price risks, they can choose the final settlement for their trades either through an RMB-denominated gold contract or spot gold delivery (Source: http://www.cgse.com.hk).

The launch of RMB Kilobar Gold has attracted considerable attention from both local and international financial communities. It is believed that RMB gold product shall play an important role in the Hong Kong offshore RMB market (CNH), which is now fully-deliverable and allows foreign investors and corporations to have unrestricted access to China's currency via cash accounts, FX, bonds, equities and through cross-border trade. Foreign companies can now invest in China with RMB obtained offshore (subject to approval) while approved onshore corporations can invest overseas using RMB.

The formation and development of CNH (the RMB offshore market in Hong Kong) is an important part of the strategy for RMB internationalization. CNH has gradually evolved because of policy measures taken by the Chinese government. It is part of the local currency offshore financial market.

1.3   The history and development of the international gold market

1.3.1   *The Imperial Power Monopoly Period (before the nineteenth century)*

Before the nineteenth century, gold was extremely rare. It symbolized the exclusive wealth and power of the Emperor. Apart from the people on Earth, gold used to be treated as a property of the Gods also. Many oblations to Gods and decorations used for protecting the Gods' images were also made of gold. Gold mines were properties of the Royals, and mining and extraction was carried out by slaves and prisoners under extremely difficult conditions. On this basis, gold had fostered the civilizations of ancient Egypt and ancient Rome.

1.3.2 The Gold Standard Period (early nineteenth century to first 30 years of the twentieth century)

In the beginning of the nineteenth century, Russia, the US, Australia, South Africa and Canada had identified mines having ore with high content of gold, and this had triggered the rapid development of gold production. In the second half of the nineteenth century, production of gold had surpassed the total production in the preceding 5,000 years. Due to the increase in gold production, there were conditions that resulted in increases in the demand of gold. Based on the rising gold production, mankind had entered into a period of Gold Standard. The Gold Standard means gold is the basis of money, is an international hard currency that can be imported and exported freely. For trading within a country, gold can be treated as a common currency. When there are imbalances in international trade, the deficit country can pay in gold to offset the difference. The Gold Standard system has three distinctive characteristics: freedom of casting, freedom of exchange and freedom to export. The Gold Standard was initiated in England in 1816. Towards the end of the nineteenth century, most major countries in the world had basically implemented the Gold Standard.

Immediately prior to the First World War (1914–19), 59 countries were following the Gold Standard. Although it was discontinued during certain periods, the Gold Standard was followed until the second decade of the twentieth century. (A number of countries, including the UK, suspended the Gold Standard during the First World War. Also, after the war a number of countries did not return to the Gold Standard and imposed capital controls. In other words, the heyday of the Gold Standard was the latter part of the nineteenth century through to WW1.) Owing to the specific situations of various countries, some of them have adopted the Gold Standard for more than a century, while some others have followed it for only a few decades.

At the beginning of the twentieth century, the outbreak of the First World War seriously impacted the Gold Standard. During the 1930s, the Great Depression in the world completely destroyed the Gold Standard. Many countries strengthened their trade controls and prohibited free trading in and import and export of gold. The open gold market lost the basis of its existence. The London bullion market, which had remained closed for 15 years, was reopened in 1954. Some countries had implemented the "National Gold Bullion Standard" or "Gold Exchange Standard" during this period, and this had greatly reduced the function of gold as a currency. This had forced gold out of the domestic currency markets. However, in international reserve assets, gold still remained as the payment mechanism of the last resort and, in turn, the world currency. During the period from 1914 to 1938, most of the gold mined in Western countries was absorbed by the central banks, and there were only limited activities in gold trading. Since then, the controls on gold had been relaxed, but the price of gold was still artificially determined by the governments. In addition, there existed trade barriers between countries and liquidity of gold was poor. The market mechanism was seriously inhibited, and the development of gold market had been seriously hampered.

1.3.3   The Bretton Woods system period (twentieth century 1940s through the early 1970s)

After a fierce debate between Great Britain and the United States in 1944, both countries reached a consensus which greatly affected the position of gold. In May of the same year, the United States invited 44 countries that had participated in the establishment of the United Nations to join a conference at Bretton Woods, New Hampshire. During the conference, the "Bretton Woods Agreement" was signed, establishing the second international monetary system after the collapse of the Gold Standard. The system called for fixed exchange rates against the USD and an unvarying dollar price of gold. And the US government took up the obligation to exchange the USD with gold at the official rate. As a result, other currencies were pegged to the USD. The USD became the currency of the world. In fact, this was a kind of new Gold Exchange Standard. In the Bretton Woods monetary system, the function of gold in circulation or its role in international reserves was reduced, and the USD became the principal player in the system. However, since gold was the final barrier providing stability to the monetary system, the price and the flow of gold were still under tight controls. Residents were prohibited from trading in gold, and the market mechanism could not play an effective role. It took more than ten years for the London gold market to recover after the implementation of the system.

1.3.4   The collapse of the Bretton Woods system

In the 1960s, the United States was in chaos because of the Vietnam War. The government deficit kept increasing and the income from international trade deteriorated. Uncontrollable inflation occurred in the US and the credit of the USD was greatly impacted. During the same period, European countries had started recovering from the Second World War, and they had accumulated huge reserves of USD. Because of the high inflation in the United States, all countries as well as their markets speculated that the USD would depreciate. In order to protect their assets, gold had become the best option. As a result, many countries sold the USD and bought gold, avoiding the risk involved in holding the USD and preserving their asset values. This made it difficult for the US government to maintain the fixed exchange rate between gold and the USD.

By 1971, the gold reserves of the United States had decreased by more than 60 percent. The government was forced to abandon the policy of a fixed exchange rate between gold and the USD, and the Western countries had delinked their currencies from the USD. The price of gold was again freely determined by the market.

1.3.5   Floating Rate Period, the collapse of the Bretton Woods system

In 1976, the International Monetary Fund (IMF) approved the "Jamaica Protocol" which was amended two years later. Major contents of the protocol include (i) Gold is no longer the monetary parity value standard; (ii) Abolition of official gold price, and the International Monetary Fund will no longer intervene in the market, that is, the price of gold will be freely determined by the market; and (iii) Terminating the regulation of compulsory settlement of funds with gold.

The IMF sold one-sixth of its gold reserves, and the profit was used for establishing a preferential loan fund to help low-income countries. A new instrument, special drawing rights (SDRs), was established in lieu of gold for certain payments between members and between members and the IMF. Since President Nixon made his speech on TV in 1971 that abolished the Gold Exchange Standard and delinked the USD with gold, the price of gold has varied freely against currencies.

During 1972, the London market price of gold increased from USD 46 to USD 64/oz. Gold then fluctuated between USD 130 and USD 180 between 1974 and 1978.

In 1979, the price of crude oil exported by OPEC soared again, reaching USD 30 a barrel. This caused the gold price to rise to USD 244. On July 3rd of the same year, a famous comedy actress demanded a USD 600,000 acting fee paid in equivalent South African gold, rather than in USD. In 1979, the gold price rose to USD 500.

However, gold not being used as a currency did not result in complete withdrawal of gold from the currency field. The function of gold as a currency still remains. Many gold coins are still in circulation and have legitimate value. The change in gold price is still an effective tool for measuring the intrinsic value of a currency and for evaluating the operating status of an economy. Moreover, gold is still an important means of reserve assets. As of 2005, central banks around the world had aggregate holdings of 32,400 tonnes of gold designated as foreign exchange reserves. This was equivalent to 22 percent of the total gold production produced in the preceding two millenniums. Privately owned gold was around 24,000 tonnes. Adding up the two, the total quantity was equivalent to 37 percent of the total gold quantity in the world as of 2005. Gold is still the only recognized alternative that can replace the current currency system. The advancement of SDR is far lower than expected, and gold is the fifth most acceptable hard currency after the USD, euro, British pound and Japanese yen.

1.4 Challenges and opportunities of the Hong Kong gold market

Since the 1990s, globalization and technological advances have continuously driven world economic growth. However, the US sub-prime crisis has exposed the softer side of globalization and technological revolution, i.e. internal conflicts in existing governance structure of the world. The past 20 years have seen several economic, financial and currency crises. The Asian financial crisis in 1997, bursting of the information technology industry bubble in the late 1990s, the US sub-prime mortgage crisis (followed by the international financial crisis) and the European sovereign debt crisis occurred one after another, plunging international financial, monetary and trading systems into serious disorder.

Consequently, gold started receiving attention again. In the midst of the ongoing economic crises, gold has once again demonstrated its unique value, price patterns and multi-functionality to the world of economics and finance. Whenever economic turbulence takes place and main economic indicators such as employment, growth, inflation and asset prices move erratically, gold has been invariably and repeatedly pursued and admired. The financial crisis in 2007–08 resulted in prices of many commodities declining while gold price, in contrast, has increased steadily from around USD 600/oz in 2007 to around USD 1800/oz in 2011.

Because of the multi-faceted properties and usage of gold, its price formation mechanism is more complex than ordinary commodities. Besides the simple supply and demand mechanism, it combines the impact of spot gold commodity market supply and demand equilibrium price formation and the mechanism of gold investment in the financial market. It is difficult to distinguish the effects of each factor and, therefore, the real factors that impact gold price in a specific period need to be analyzed. Gold price is the result of the combined effect of many factors and the roles of various factors is not static, but changes in different periods.

In addition, to promote the development of a Hong Kong offshore RMB financial centre is an important measure to consolidate and strengthen Hong Kong's position as an international financial centre, which is attracting the attention of the government and industry in Hong Kong. However, in recent years the development of a Hong Kong offshore RMB centre has not been as quick as was expected. Problems and obstacles include, for example, imperfect inverse flow mechanism, narrow investment channels, Mainland financial system constraints, and so on. The gold market has the advantages of a Hong Kong financial market, compared to other financial centres, and should play an increasingly important role during the process of promoting construction of the Hong Kong offshore RMB centre.

At the same time, observing the trend of RMB internationalization, how the Hong Kong financial market and gold market can seek their respective development opportunities while facilitating RMB internationalization are important questions.

1.5   Significance of the topic

It seems that throughout the history of the monetary system, the outbreak of every large-scale financial crisis has resulted in profound changes in the international monetary system. The Great Depression of the 1930s played a significant role in the configuration of what is known as the Bretton Woods system. The early 1970s oil crisis led to global inflation and exchange rate volatility, accelerating the pace of European monetary integration and unification. The 2007 US sub-prime mortgage crisis triggered a global financial crisis, and international monetary system reform has once again been put on the agenda. Reform of the international monetary system has been a continuous process and yet there is no consensus so far. Gold is once again being viewed as a hedge asset. In the international monetary system, gold may play a more important role in coming years. Research on the structure of the gold market and its role in global financial markets may help comprehend the future trends.

On the practical side, gold prices are affected by mining, recycling, industrial usage and jewelry consumption, besides investment demand. If the factors that impact price are well analyzed, significant business benefits can be obtained by those dealing in gold.

At present, confidence in currencies is decreasing, and this lack of confidence is beginning to spread to all associated financial assets. However, due to the special price formation mechanism, gold has become a powerful tool to hedge against many different kinds of financial risks. In this situation, research on the gold price formation mechanism and gold market operation mechanism can help the process of hedging against volatility of several investment products and for reducing investment risks.

Exchange rate of the USD plays a critical role in gold price movements. With the reform of the exchange rate formation mechanism, independence of the gold RMB offer is expected to be strengthened. Linkage between the Mainland gold market and foreign exchange market may also get enhanced. Arbitrage opportunities between gold price in USDs and RMBs may emerge in not too distant a future.

The internationalization of the currency cannot be implemented without the offshore financial markets. Before becoming a completely and freely convertible currency, the establishment of offshore financial centres should provide the country's financial institutions, in particular domestic commercial banks, opportunities to adapt to international financial transactions, and then transiting to a convertible currency. The RMB offshore financial centre will greatly promote the process of RMB internationalization.

In Hong Kong, the geographic location, political environment, institutional arrangements of financial markets, legal system and the accumulation of talent and experience in being an offshore market for other currencies, provide sufficient conditions for the development of the offshore RMB market. The Hong Kong financial industry should seize this opportunity.

Unit of analysis and stakeholders:

This book discusses gold markets around the world and focuses on the Hong Kong and China gold market.

The stakeholders of this research are:

a   Gold mining enterprises: can gain better understanding of price formation mechanisms which can help reduce pricing risk.
b   Companies who use gold: can hedge against price and cost risk more efficiently.
c   Investors in the gold market: can analyze the trend of gold price.
d   Gold exchanges: can design better products to offer to investors.
e   Gold market regulators: can better monitor trade and transactions.
f   Government: can facilitate RMB internationalization and the Hong Kong offshore market.

2 Gold property and gold market mechanism

One of the major purposes of this chapter is to identify the key factors that can shape the dynamics of gold prices, by drawing on relevant literature (see Figure 2.1). These factors are considered in the development of the theoretical model that explains the recent price dynamics experienced and to predict whether it is the right time to enter the Mainland market. The other purpose is to summarize extant research related to currency internationalization and arbitrage; we should try to analyze historical experience of other currencies from the literature.

Currently, the main institutions conducting research on the gold market include Thomson Reuters GFMS (formerly known as GFMS or Gold Fields Mineral Services Limited), World Gold Council (WGC) and gold sections of various exchanges and reserves departments in major countries' central banks. The research can be divided into two main topics: one is study of the market itself, such as the operational mechanism of the market, gold derivatives, gold commodity in banking business, the world's gold supply and demand, etc.; and the second is the study of gold price, such as factors that influence gold price, volatility of gold price, correlation of gold and other financial assets and so on.

My research will mainly focus on the second topic. I introduce the literature on this grouping herein below.

The literature reviewed in this part should be divided into three major sections. First, some literature has examined inherent properties of gold and the gold market in history and at present. These help the readers understand the gold market from a more macro perspective and help some interesting phenomenon that may not have been well discussed get noticed. Then I review literature about the kernel of the market – gold price. In this part, characteristics and special dynamics of gold price are discussed. And then literature about factors that influence gold price is reviewed. Some have described the mechanism of gold price in special periods but further works can continue to examine the situation in new periods and areas. In the third part, another important topic of gold market, relationship with other financial assets, is also reviewed. Discussion of this topic can provide some support for the special properties of the gold market and gold price.

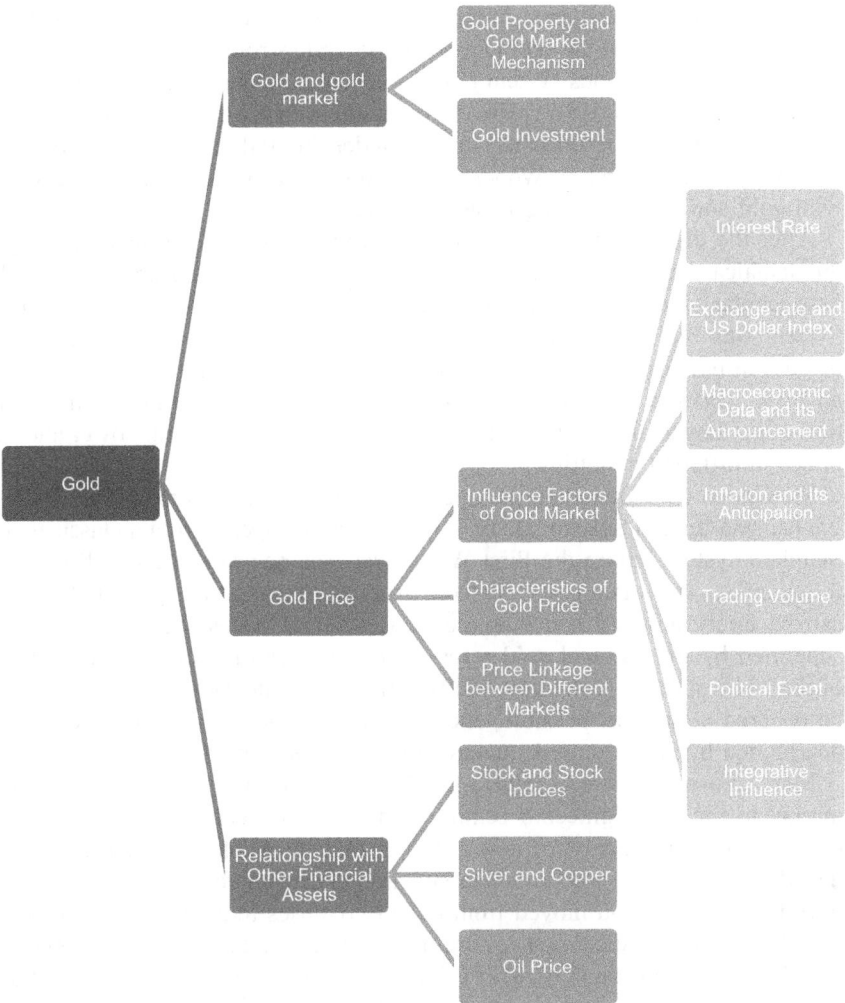

Figure 2.1 Literature grouping

2.1 Gold product

The most important and useful source of information on this topic is the annual "Gold Survey" now published by Thomson Reuters GFMS to sum up the global gold market development and changes over the past year. This publication was first produced in 1967.

As early as 1938, British economist Paul Einzig (1938) had investigated gold forward markets when researching exchange rate determination theories.

In 1960, Belgian-American economist Robert Triffin first identified the Triffin dilemma (or Triffin paradox), that is, when a national currency also serves as an international reserve currency, there can be conflicts of interest between short-term domestic and long-term international economic objectives (Triffin, 1960). The theory articulated the problems with the USD's role as the reserve currency under the Bretton Woods system (Gold Exchange Standard, fixed exchange rate between gold and the USD). The embarrassment of the USD and the instability of the system were destined to happen, under the Triffin dilemma. Along with the collapse of the Bretton Woods system, the USD lost the fixed exchange rate with gold and its dominating position in the world.

After the disintegration of the Bretton Woods system, IMF members signed the Jamaica Agreement that formalized the floating rate system in 1976. The first scholar to provide a full and detailed description of gold in the Jamaica System was Italian economist Alberto Quadrio-Curzio. He elaborated the role of gold in nations and the international monetary system in his book, *The gold problem: Economic perspectives* (1983). He proved the value-maintaining function of gold by analyzing the volatility of currencies' purchasing power and gold prices in different countries.

McKinnon (1993) studied the operating rules of the gold market under Gold Exchange Standard System. He believed that stable operation of a classical Gold Standard system before the First World War was dependent upon all countries following six important rules: (1) fixed official price of gold is persisted; (2) gold can be freely inputted and outputted; (3) all bank notes and coins should be supported by the national gold reserves; (4) the central bank should be lender of the last resort; (5) if the first rule is temporarily interrupted, the parity should be restored as soon as possible; and (6) nominal price should be endogenously determined by supply and demand of the world gold market.

Marcuzzo and Rosselli (1987) studied available data on the London gold market in the late eighteenth century and early nineteenth century. The paper presented the formulation of new gold points and modeled the market behavior that determined gold price movements. A mechanism for gold movement was introduced where gold moved from deficit countries to surplus countries in the world, in order to adjust and restore the balance of international payments.

Salant and Henderson (1978) considered that the mismatch between gold price volatility and the depletion resource model was mainly because governments adjusted the amount of gold reserves in the market according to demand. Anticipation of official adjustment should increase the uncertainty of the market. The effects of anticipation of government sales policies on the real price of gold are analyzed in this research on gold market.

Aggarwal and Soenen (1988) first researched the nature of the gold market, and discussed commodity and financial properties of gold with emphasis on the financial attributes. Then they researched the efficiency of the gold market and contested the claim that the gold market in the US was efficient and that the US gold futures market was able to discover the price of gold appropriately and efficiently.

Bertus and Stanhouse (2001) considered that the supply of gold should not fluctuate widely because the amount of gold hoard is much greater than the annual gold production, about 50 times, and the production is not seasonal. Also, the 24 hour gold trading around the world can help ensure that any departure of the price from the fundamental value of gold can be adjusted quickly. With this character, the gold market should have little opportunity to produce a speculative bubble. But in the 1980s and 1990 to 1993, the gold spot market and the futures market underwent significant changes. The authors used a dynamic factor analysis method to prove whether there was a price bubble. The result showed that based on the traditional significance level, it had failed to show a bubble. But if the confidence interval is relaxed, bubbles may occur with major events in history.

Burton (2005) investigated the structure of the gold industry and compared the structure in the 1980s and that in the new century. The paper explained what had happened in the intervening years by analyzing the main industry forces in the context of countries, companies, consolidation and costs which indicated that the price should remain approximately at an average of USD 250/oz.

2.2   Gold investment

Blose (1996) presented a model for estimating theoretical elasticity of gold price for determining the economics of investments in mining companies by mutual funds and concluded that gold can be a good tool to hedge against systemic risk.

Scott-Ram (2001) compared adding bonds and gold to a portfolio for the long-term and found gold investment more effective in spreading of risks.

Richard Michaud et al. (2006) concluded in their WGC report that adding 2–4 percent gold to the portfolio offers clear strategic benefits.

Hillier et al. (2006) reported that precious metals including gold have hedging capability, particularly during periods of "abnormal" stock market volatility, by analyzing data from 1976 to 2004.

Dempster (2008) researched gold price fluctuation under the background of a recessionary environment in 2008. The study reported that the price of gold has its own unique driving mechanism; there is no negative impact on the price of gold during economic downturns. So adding a certain proportion of gold investment into an investment portfolio can be good for avoiding or reducing risk in recessionary periods.

Baur and McDermott (2010) examined the role of gold in the financial system and considered gold more stable in terms of profitability, relative to stocks and bonds and other financial instruments. Especially, the gold market can reduce the extreme negative impact of the financial system and effectively stabilize financial markets.

Colin Lawrence (2003) concluded that although the gold market is affected by gold supply and demand, its return is independent of business cycles. The main macroeconomic variables such as GDP growth, inflation, interest rates and others have impact on the price of gold, but have no significant impact on gold investment returns; there is no statistical relationship between them. But returns of other financial assets, such as the Dow Jones Industrial Average, S&P 500 index and ten-year bonds are significantly associated with these macroeconomic variables.

2.3   Gold price – influence factors of gold price

2.3.1   Interest rates as important influence factors

Koutsoyiannis (1983) indicated that there is a very strong correlation between gold price and US interest rates, after analyzing a sample of 316 daily observations of gold bullion prices over the period from January 1980 to March 1981.

Fortune (1987) presented that anticipation of an increase in interest rate causes a negative adjustment of the price of gold, after estimating a model with quarterly data for the United States over the period from the third quarter of 1973 to the second quarter of 1980.

Harmston (1998) researched data of US Treasury bills and long-term government bonds from 1896 to 1996 and showed negative correlation between the trend of gold price movements and value of T-bills and other long-term bonds.

Cai et al. (2001) analyzed 5-minute returns of COMEX gold futures for a sample period of 4 years from 1994 to 1997 and believed that in usual cases, changes in gold prices were highly affected by changes in interest rates.

Monroe and Cohn (1986) regarded short-term Treasury bill rates as opportunity cost of gold storage. The deviations of implied gold interest rates from T-bill interest rates allow traders to earn profits from speculating on changes in the difference between the two rates, i.e. gold futures prices respond to term structure of interest rates for balancing. The article researched gold and Treasury bill futures prices on the Chicago Mercantile Exchange from March 1976 to July 1982 and the results indicated that profits were consistently available if one took a position based on a return toward equilibrium.

2.3.2   Exchange rate and US dollar index

Sherman (1983) stated that the log of the gold price was negatively related to the US trade-weighted exchange rate.

Dooley et al. (1995) used end-of-month exchange rates between the USD and three other major currencies (the pound sterling, the Japanese yen and the Deutsche mark) over the period 1976–90. They found gold price movements to have explanatory power with respect to exchange rate movements.

Larry and Fabio (1996) applied theoretical models and provided empirical evidence to prove that since the collapse of Bretton Woods, the international monetary system, floating exchange rates among major currencies had been a major source of price instability in the world gold market and, as the world gold market is dominated by the European currency bloc, appreciation or depreciation of European currencies has strong effects on the price of gold in other currencies.

Capie et al. (2004) explored gold price and exchange rates of various currencies against the USD from 1971 until June 2002 and found the USD gold price moving in the opposite direction from the USD and gold was the most effective tool to prevent the losses caused by the depreciation of the USD.

Tully and Lucey (2007) explored the impact of main macroeconomic factors on the price of gold by analyzing monthly observations of gold, both cash and futures prices and a set of macroeconomic variables over the 1984–2003 period. The article focused on data of 1987 and 2003 and claimed that the USD is the main factor affecting the price of gold. In many cases, the USD price is the only macroeconomic factor affecting price of gold. Gold can stand against devaluation of the USD. Although they presented very different conclusions because of different sampling methods, they summarized the major factors that influence gold prices and presented a relatively uniform explanation.

Sjaastad (2008) examined theoretical and empirical relationships between major exchange rates and the price of gold using forecast error data and considered that uncertainty of exchange rates among major currencies is the major source of gold price instability. As the world's major gold market is primarily dominated by the USD bloc, appreciation or depreciation of the USD has strong effects on the price of gold in other currencies. However, results of this study are very different from previous studies. The results show that during the 1980s, the world gold market was dominated by the European currency bloc; in the 1990s and the early years of the current century, the market was co-led by the USD area and Japan. Major gold producers (Australia, South Africa and Russia) have little influence on world gold price and gold is no longer a hedge against inflation.

2.3.3   Macroeconomic data and its announcement

Fama and French (1988) studied the impact of business cycles and macroeconomic environment on gold futures prices and carried out an empirical research comparing volatility of gold futures contracts in the United States and Japan. They showed that volatility of gold futures contracts in the United States was larger than Japan, and indicated that this difference may be due to differences in information transmission.

Tandon and Urich (1987) analyzed the effect of money supply and inflation announcements on gold price for the period from July 1977 to December 1982. They found that unanticipated changes in money supply had little effect on gold price before October 1979 but were negatively related to the changes in gold prices after that date and unanticipated increases in PPI were positively related to gold prices while CPI had no effect.

Christie-David et al. (2000) documented response of gold futures prices to monthly macroeconomic news releases from 3 January 1992 to 29 December 1995. They reported that gold prices strongly reacted to release of CPI, GDP, PPI and unemployment rate but weakly responded to release of data of federal deficit.

Cai et al. (2001) examined the impact of 23 regularly released macroeconomic announcements in the United States on gold price covering the period from 1994 to 1997 and found four significant effects on volatility of the gold market. Employment reports, followed by GDP, CPI and personal income were found to be the most important announcements for the gold market.

Lauterbach and Monroe (1989) tried to analyze effects of information and noise on intraday gold futures returns by examining gold futures transaction prices in Chicago Mercantile Exchange from August 1978 to February 1979. The results showed that the distribution of futures returns varies from trading to non-trading hours. The variability of returns is higher during the first half hour and the last hour of the trading day; it is especially high during the first few minutes. Possible causes include information levels of public and private as well as noise trading.

Spyrou (2006) analyzed the CMX Gold 100 Oz Futures contract data covering the period between July 1990 and April 2005 (3,850 observations for each contract) to research the impact of arrival of unobservable information on investors behaviors (overreact or underreact to the information) in gold futures market. The result showed that investors in the gold futures contract appear to react efficiently to this type of information.

2.3.4   *Inflation and its anticipation*

Jastram (1977) studied gold price movements from 1560 to 1976 and reached the conclusion that the gold market performs better in deflationary periods than in inflationary periods.

Sherman (1983) claimed that gold price had a positive correlation with level of inflation, irrespective of whether inflation anticipation exists or not.

Baker and Van-Tassel (1985) examined monthly gold price in 1973–84 and concluded that if inflation is expected, the percentage increase in gold price will be greater than the percentage fall in the dollar's value.

Mahdavi and Zhou (1997) compared efficiency of gold and commodities prices as leading indicators of inflation. They analyzed quarterly data from 1970 to 1994, but found no evidence of a correlation between CPI and London price of gold. They claimed that commodity prices may be a better leading indicator of CPI.

Stephen Harmston (1998) studied cases from the United States, Britain, France, Germany and Japan. The article believed that if measured with real purchasing power, gold maintains its value in the long term. Although the price of gold keeps fluctuating, it always returns to purchasing power parity. During war, commodities prices rise faster, but the mobility, acceptability and portability of gold are more important than the exchange rate based upon commodities in these times. After comparing gold supply and demand basis in 1971, the article considered that gold's purchasing power increased in the 1970s partly because of release of the long-term fixed gold price and partly because of inflation. After 1971, the world-wide increase in demand for gold pushed up gold price relative to other commodities, while during the late twentieth century, the increase in gold supply made the purchasing power of gold return to its historical average.

Taylor (1998) researched the relationship between inflation and prices of precious metals (gold, silver and platinum) from 1914 to 1937 and 1968 to 1996 and found that in the period before 1939 and around the second OPEC oil shock in 1979, the ability of gold to hedge inflation was concentrated in both short and long-term.

Laurence E. Blose (2000) examined unexpected changes in CPI and PPI in the United States from March 1988 to February 1999 and found that both gold prices and gold futures prices were not affected by unexpected changes in these two indexes.

In the book *A Historical Review: American Gold Market*, Kennedy (2002) concluded that gold rose slower than other goods in early inflation, but after the initial period, price of gold rose much faster than other commodities, i.e. the increase is much larger than the inflation rate, which can increase its value and provide hedge against inflation.

Ranson and Wainwright (2005) believed gold price to be a leading indicator of inflation and bond market, and it can act as a strong predictor of inflation and short-term and long-term nominal interest rates. Because gold price moves in the same direction as inflation, it is a good tool to build a portfolio to prevent loss because of inflation.

Laurence E. Blose (2010) analyzed monthly CPI announcements from March 1988 to February 2008 and showed that surprises in the CPI do not affect gold spot prices and investors cannot determine market inflation expectations by examining the price of gold.

2.3.5 Trading volume

Bhar and Hamori (2004) examined the mode of information transmission between the percentage price change and the trading volume in gold futures contracts using daily data from 1990 to 2000. The result showed strong causality in variance from percentage price change to trading volume with a lag of ten days. This result is just the opposite of the result of agricultural and crude oil futures. The possible causes may be due to the natural properties of gold as a commodity and its special hedging role when the stock market performance is bad.

Jonathan and Brian (2007) applied the GARCH model to research gold futures contract on the Chicago Board of Trade (CBOT) using intraday data from 1999 to 2005. They explored the characteristics of price volatility and concluded that effect of trading volume on the price of gold futures was almost negligible.

2.3.6 Political event

Barkoulas et al. (2008) found that during an unexpected geopolitical event, in the short run, gold price rises while the gold mining companies' stock prices fall.

Melvin and Sultan (1990) examined the impact of South African political unrest on the gold market. The article demonstrated that there was no particularly significant relationship between South Africa's political unrest and gold futures premium.

Extant literature has not discussed trading volume and political events adequately but I have tried to research the influence of these two factors as some exploratory work.

2.3.7   Integrative influence

Levin and Wright (2006) analyzed data from January 1976 to August 2005 to test factors that influence short-run and long-run movements in gold price and provided some important conclusions. There is long-term, fixed and positive correlation between gold price and US price level; 1 percent change in US general price level causes 1 percent increase in price of gold, so gold can be regarded as long-term inflation hedging tool. Sudden events impact the long-term relationship but there is a slow reversing process. However, world income and volatility of global inflation have no significant correlation with gold price. In the short-run, there are several factors that significantly influence the price of gold: US inflation and its volatility; credit risk has a positive effect on the price of gold; increase of the USD index and actual interest rate reduce the price of gold; and world inflation and its volatility, global income and gold risk have no significant effect. In addition, the authors believe political risk of oil-producing countries and financial crises are important factors that impact gold price fluctuations.

Baker and Van-Tassel (1985) conducted regression tests of monthly gold prices during 1973–84 by exchange rates, interest rates, inflation rates, changes in commodities prices and speculation and concluded that there are significant relationships between gold price and the USD, inflation and changes in commodities prices, but there is no significant relationship between gold price and interest rates.

Wang Chengbiao et al. (2007) considered that the price of gold is affected by the world economic situation, USD exchange rate, inflation, international situation, international trade, crude oil prices, foreign exchange policy, the stock market, fiscal deficits, unrest, war and other factors. Most of these factors are gray. So they established a model to simulate and predict international price of gold and improve prediction accuracy.

2.4   Characteristics of gold price

Tschoegl (1980) and Solt and Swanson (1981) both found that fluctuations in price of gold are time-dependent.

Cheung and Lai (1993) believed that after 1979, generation of long-term memory in the gold market was mainly due to several factors and along with the disappearance of these factors, long-term memory will vanish.

Terence Mills (2004) examined a statistical description of daily gold price on the London market from 1971 to 2002 and found that the price of gold had significant sharp peaks and heavy tails and was non-normal.

Edel (2006) applied the GARCH-M model to research the relationship of benefits and risks of the gold market using daily gold futures and spot price in the New York Mercantile Exchange from 1982 to 2002. The article found that the price and expected return did not include a very significant risk premium.

2.5   Price linkage between different markets

Dhillon et al. (1997) compared volatility of gold futures contracts in the United States and Japan. They indicated that the volatility of gold futures contracts in the US market to be stronger than Japan and this difference may be due to the different information transmissions.

Chow (2001) applied the AR-GARCH model to research information links for the US–Japan cross-market gold futures contracts. The results found that price information transmits very fast between the two markets and the US market is in a dominant position.

Xu and Fung (2005) researched metals futures in the New York and Tokyo Exchanges and analyzed the relationship between them. They found strong price spillover effects between the two markets' gold futures and information transmission was very fast.

2.6   Relationship with other financial assets

2.6.1   Stocks and stock indices

Ivanova and Ausloos (1999) collected daily data from February 1991 to May 1997 to analyze relationships among price of gold, Dow Jones industrial average and the exchange rate between the Bulgarian lev and USD.

Graham Smith (2001) studied the relationship between gold prices in USDs and several major US stock indices covering the period from January 1991 to October 2001. The result showed that coefficient of correlation between gold price volatility and major industrial countries' stock indexes was not significantly zero. There are weak negative correlations between them. They have different fluctuations and fluctuations of stock price have a slight impact on gold price but they are not the direct cause. Smith (2002) obtained very similar conclusions on examining the case of Europe and Japan for the same period.

A research report of the World Gold Council (Kennedy Steven C. from WGC, 2002) indicated that the price of gold has negative correlations not only with stock indices, but also with most other financial assets' prices.

Pulvermacher (2003) studied the relationship between gold and stock in US and UK stock markets covering the period from 1973 to 2003. The article pointed out that although there is a clear reverse relationship between gold and stock volatility, there does not exist a clear correlation between returns of gold and stock index.

2.6.2   Silver and copper

Urich (2000) discussed the stochastic structure of metals futures prices. He first proposed a fixed multi-factors model, then extended the model allowing changes of the time, and applied the model to three different metals futures prices (copper, gold and silver). The results showed that the correlations among them had different patterns of fluctuations.

Adrangi et al. (2000) used 15 minute intraday data of gold and silver futures contract in New York Mercantile Exchange from 27 December 1993 to 30 December 1995. They analyzed the price discovery process among the strategically linked gold and silver futures contracts and examined the role of the inter-market spread in their price dynamics. The data showed that silver futures contracts bear most of the responsibility for convergence spread between gold and silver and the strategic link between price of gold and silver futures contracts.

Ciner (2001) analyzed the long-term price trend of gold and silver futures contracts in the Japan Commodity Exchange. The results showed that the stable relationship between gold and silver prices disappeared in the 1990s. Then the article discussed the root causes and effects of the phenomenon.

Liu and Chou (2003) examined gold and silver futures with spot data from January 1983 to July 1995 to analyze parity relationships between gold and silver in futures and spot markets. The results revealed that the two markets showed a long memory and slow adjustment process; in other words, they have a long-term equilibrium relationship.

2.6.3   Oil price

Baffes (2007) empirically proved that prices of precious metals have significant impact on crude oil prices.

Malliaris and Malliaris (2009) collected data of gold, oil and the euro from 4 January 2000 to 31 December 2007. They applied time series analysis and a neural network model to analyze the relationships among the three variables. The results of the time series analysis indicated that there exist short- and long-term relationships: oil follows the change of gold price, euro and oil have equivalent influence and the relationship between gold and the euro is the weakest. Neural network results showed that the influence of oil on gold is larger than gold on oil and influence of gold on euro is larger and faster than euro on gold.

Zhang et al. (2009) proposed the establishment of the Bozeman neural network model based on wavelet transform using gold prices and oil price data. The research indicated there is a significant correlation between the two prices which affect each other.

Literature on the relationship of gold with other financial assets, most important and emphasized assets, stocks and stock indices, oil, silver and copper was reviewed. Similar to the factor of inflation, the results are diversiform. I will follow the discussion and do more research on recent market trends in Hong Kong and China.

3 In-depth interview on topical issues in the gold market of Hong Kong

The interviews were conducted in the form of semi-structured conversations. Fifteen participants took part in this survey, all of whom are related to the gold trading business to various extents. Among these fifteen participants, six of them are related to gold trading business, while another five of them are senior executives of asset management companies. Another three of them are senior executives of major gold consuming companies (jewelry chain stores) and one is a professor of finance from a university in Hong Kong. Apart from having business in Hong Kong, some of these participants have trading relationships with entities abroad, directly and indirectly.

The participants in this survey are the senior executives and decision makers of their organizations, and their job titles include, but are not limited to, chairman, managing director, director, chief strategist and senior associate.

These fifteen participants were invited for interviews at their offices, and they were asked ten pre-determined questions. These ten questions are grouped in seven major categories, and their responses are aggregated and summarized in the following sections.

3.1 Major factors affecting the price of gold

Most views suggest that the price of gold adopts USD as the currency for trading and clearance. Under normal situations, in terms of economic development, capital and supply and demand of gold, etc., the price of gold will in general change in the reverse direction against the value of the USD. The Chief Dealer in the Forex & Bullion Dealing section of HSBC points out that statistically the correlation between the two is -0.85. Furthermore, the price of gold is affected by conventional factors such as the price of crude oil, international political environment, inflation and interest rate of USD, etc. Here, gold functions as a safeguard for the value of assets against inflation and acts as a shelter for protecting the assets against major political incidents and wars.

However, some views suggest that the local political effects on gold prices are relatively temporary. The Non-Executive Director of Chow Tai Fook said that apart from other investment vehicles such as stocks and bonds, investment in gold cannot enjoy benefits resulting from dividends or interest. As a result, the appetite for investing in gold is diminished when the interest rate increases or when the stock market soars, as the opportunity cost of investing in gold is comparatively high. On the other hand, higher interest rates mean higher financial costs for the gold miners. This reduces the interest of the gold miners in mining for more gold and thus it affects the supply of gold in the market.

3.2   Change in factors affecting the price of gold

The CEO of Chancellor Precious Metals suggests that the conventional price of gold is also affected with respect to different periods of time. Interest rates have been at historical lows in recent years and therefore their impact on the price of gold is relatively insignificant. A few decades ago, when the USD was strong in comparison to other currencies and interest rates of the USD were high, gold used to be an effective tool to protect asset values against inflation. The price of gold was comparatively sensitive to the interest rate in those days, but recently their impact has become relatively less significant. The markets of newly developed countries rely less on crude oil and therefore the influence of crude oil on gold price is also diminishing.

When the price of crude oil surged to USD 140 per barrel, the price of gold had reached USD 1,800/oz. Thereafter, the price of crude oil had gone down by 70 percent, but the price of gold declined only 50 percent because of other factors supporting the price. In recent years, the price of gold has been predominantly affected by the quantitative easing policy of the United States, which in turns affects the value of the USD and the fluctuation in the stock market caused by the market's confidence level in the USD future.

The entire world financial system has been stepping out from the financial crisis of the United States and Europe in the last couple of years. Nevertheless, the market is still focusing on the excessive borrowing issues in the United States, Europe and Japan and there are still high potential risks in credits related to the USD. This could be one of the major reasons for the high volatility of gold price in the near future. If systemic risks occur in the U.S. bond market, gold will become the favorite tool for asset protection and be welcomed by the investors.

In addition, the responsible officer of Po Sang Financial Investment Services Company Limited suggests that apart from the conventional factors, the cost of mining, recycling and the emergence of new technology are also affecting the price of gold. Unlike crude oil, products made of gold can be recycled for reuse, such as gold ornaments. However, the crude oil product can hardly be recycled. As a result, the amount of gold explored through mining is a process of accumulation from a global perspective.

With the improvement of the quality of life, the demand for gold is different in different locations. For example, in Mainland China, more people use gold-plated jewelry instead of real gold jewelry for wedding purposes, mainly because of the technological advancement in producing gold-plated jewelries. Furthermore, consumer preferences have changed their interests from pursuing gold jewelries to some high-end precious stone jewelries.

3.3 Specific factors affecting the gold market in Hong Kong

The Chief Dealer in Forex & Bullion Dealing of HSBC said that Hong Kong is one of the ten largest gold trading markets in the world, and its position is just second to London and New York. Its distinctive position in this part of the world provides Hong Kong a unique development advantage over other gold trading markets. Owing to its historical background, London is the most active and the largest spot gold trading centre. South Africa is a key export country and the United Kingdom has its own predominance. Much of gold is transported to London for pricing and trading. Moreover, most of the participants of the London gold market are institutional investors instead of individual investors.

New York dominates trading in gold futures and its history is comparatively long and successful. There are relatively comprehensive regulations that govern gold trading and therefore major banks and gold trading companies select the New York futures exchange (the Comex) as the trading platform for arbitrage. Nowadays in New York, there are more and more investors from Mainland China.

3.4 Unique location advantage of Hong Kong

The Permanent Honorary President of the Chinese Gold & Silver Exchange Society said that because of different but complementary time zones, Hong Kong, London and New York share the global financial markets by allowing investors to participate in the world markets round the clock. For gold trading, the two largest physical gold markets in the world are China and India, and both are located in Asia. Strong demand for spot gold causes volatility in gold price, and volatility of gold price triggers spot gold trading in Hong Kong.

Major gold production countries include South Africa, Australia, Russia and China. London is a traditional gold trading and pricing centre. Hong Kong is backed by the two major gold consumption countries, China and India, which consume more than 50 percent of gold globally. Nevertheless, Hong Kong is relatively close to other gold production countries as well and therefore Hong Kong has a comparatively higher potential to be a gold trading centre.

On the other hand, Hong Kong has its unique locational advantages in the East Asian region. Singapore is at the southern end of East Asia and Japan is in the northern part. Hong Kong, however, is in the central part of East Asia, and it takes less than four hours of flight from Hong Kong to any major financial centre in East Asia. More importantly, Hong Kong has strong backup from Mainland China, who has enjoyed substantial growth of economy in the last couple of decades. As the bridge to China, Hong Kong has its natural geographical advantages and can benefit from the high-speed economic development of Mainland China. The free market for international currencies, the simple tax laws, the comprehensive regulations of trading and the fair jurisdiction of Hong Kong provide a very important international platform for the economic development of China.

Taking into account the gold trading market history of Hong Kong, a large group of investors from Shanghai were the major gold trading participants after

World War II. Since the economy of China was opened up in 1978, more investors from the Mainland have participated in gold trading and investment. According to the data from the Hong Kong Census and Statistics Department, Hong Kong imported 800 tonnes of gold in 2013, most of which was re-exported to the Mainland.

3.5   Mature market economy of Hong Kong

Most of the interviewees mentioned that being the world's most liberal economic entity and having the comprehensive and effective judicial system in place, together with the unique "one-country-two-system" administrative advantages, Hong Kong has fostered a mature market economy system. Based on this system, the ownership of asset is clearly defined. The classification, registration, transaction and security of stock, asset, bond and other financial derivatives are all well protected under the legal system. This has greatly enhanced the confidence of the investors towards the Hong Kong market. Furthermore, Hong Kong provides a level playing field for both local and international investors, and all investors are able to compete on a fair basis. Foreign currencies are free to move in and out from the Hong Kong market, whilst new financial products emerge continuously.

The financial market of Hong Kong has a very high degree of transparency. Investors can quickly obtain financial information from all over the world conveniently and securely. As the transaction volume of Hong Kong financial market is huge, transaction levies are low and the market is comparatively flexible, which greatly reduces the cost of financial transactions in Hong Kong.

At the same time, some famous gold jewelry suppliers, such as Chow Sang Sang and Chow Tai Fook, who are also members of the Hong Kong Gold and Silver Exchange, have created brand names for Hong Kong gold products, and these gold products are well recognized by consumers and investors from Mainland China, Taiwan, Malaysia, etc. This has also provided confidence in developing Hong Kong into a gold trading centre.

Apart from gold futures trading, traditionally the vast volume of transaction in spot gold is another trading characteristic of Hong Kong. Gold products of Hong Kong constitute the first commercial product backed by the Trade Description Ordinance, providing the international market protection on quality and reputation.

3.6   Development of the RMB gold market in Hong Kong

The responsible officer of Po Sang Financial Investment Services Company Limited said that the most important part of the RMB Gold Market is "RMB". At present, most of the gold trading is denominated in USD while traditionally the trading in CGSE has been denominated in HKD. With the internationalization of RMB, there is huge potential for the RMB gold trading market.

On 17 October 2011, the Hong Kong Gold and Silver Exchange CGSE launched the first in the world "RMB denominated Gold Kilobar". RMB denominated gold means spot gold products which have the same specifications as those in the Shanghai Gold Exchange. These spot gold products are 9999 Kilobar produced by recognized refineries of the CGSE, and they can be traded via the electronic trading system of the CGSE. The CGSE has 104 years of history in gold trading. It experienced rapid growth after World War II, particularly in the 1980s and the 1990s. Moreover, it has successfully introduced the participation of foreign gold traders and international investors.

On one hand, the development of RMB gold will provide convenience to the customers such that it avoids the complicated formalities and the loss due to exchange rate. On the other hand, it can increase the size of the capital pool of offshore RMB market.

The Chief Dealer in Forex & Bullion Dealing of HSBC said that China has surpassed the United States as the largest trading volume country for importer and exporter in the world. In 2013, RMB became the seventh largest currency in the world and the second largest currency for trade financing. By 2020, it is anticipated that the RMB will become the third largest trading settlement currency. China has continuously developed free trade zones with adjacent and other countries, such as Japan, Singapore, Malaysia and Indonesia in ASEAN, Switzerland in Europe and Brazil in South America. With the implementation of RMB as the settlement currency for China trades since 2009, in 2012 about 10 percent of China trades used RMB as the settlement currency, and this figure increased to 16 percent in 2013.

Using crude oil as the example, Russia and the Middle East are the major sources of crude oil import for China. As the crude oil is traded using the USD, this reduces the geopolitical risk. On the other hand, Russia and Australia are the major countries exporting gold. If the settlement is made in RMB, it will help these two countries in procuring daily supplies and consumables they import from China.

With the development of RMB as the settlement currency for bulk cargos, the future of RMB gold is also considered optimistic in general. China is the largest gold producing country, while at the same time it is the largest consuming country as well. Last year, China consumed 1,160 tonnes of gold, while the total gold consumption in the whole world is about 4,000 tonnes. That means China consumed around 30 percent of gold worldwide and the market for RMB gold is enormous.

3.7 Advantages of developing the RMB gold market in Hong Kong

As the Senior Vice President, Head of Central Dealing of Sun Hung Kai Financial said, Hong Kong has a strong competitive edge in developing RMB denominated gold. The gold market in Hong Kong is relatively mature. There are well-developed associated facilities which facilitate the trading of gold in most of the foreign currencies. As an international financial centre, investors are confident when participating in the gold market of Hong Kong. Moreover, the century-old gold trading history of Hong Kong has provided a strong foundation for the development of RMB denominated gold. Here, CGSE has played a very important role.

The Permanent Honorary President of the Chinese Gold & Silver Exchange Society said that with Hong Kong being the most important offshore centre for RMB trading in the world, the launching of RMB denominated gold products will further strengthen Hong Kong's position. The RMB savings in Hong Kong have been increasing in recent years and the total amount is close to RMB 1 trillion. The formation of a large RMB capital pool in Hong Kong also provides a good foundation for RMB gold trading, while RMB gold trading has become an important investment channel for the RMB denominated capital in Hong Kong. In turn, this will increase the flow of offshore RMB currency.

As the largest RMB offshore centre in the world, Hong Kong has accumulated around 70 percent of the offshore RMB funds in the world. Shanghai is the onshore RMB gold trading market, while Hong Kong is the offshore RMB gold trading market. Both markets have great potential to process the hedging transactions.

The Managing Director of Pegasus Fund Managers Ltd. summarized that the internationalization of RMB can be broadly classified into three stages: (i) as the major settlement currency in the world trading market; (ii) as the investment vehicle providing various RMB denoted investment products; (iii) as the reserve currency (on one hand it becomes the central reserve currency for central banks of other countries and on the other hand it is accepted by general investors such that it becomes the saving currency of individual investors).

An Associate Professor of Department of Finance & Decision Sciences of Hong Kong Baptist University stated that at present, internationalization of RMB is at its preliminary stage and it is approaching the direction of becoming the world's major settlement currency for trading. The People's Bank of China has increased the daily volatility of the RMB exchange rate from 0.5 percent to 1 percent, and subsequently to 2 percent. Moreover, the direct share-trading between Hong Kong and Shanghai stock exchanges will further enhance the investment function of the offshore RMB.

On the other hand, internationalization of the RMB will start from the nearby countries and markets such as Hong Kong, Macau and Taiwan, and then further penetrate other Asian and ASEAN countries. The final stage of the development will be penetration into markets in Europe and America, while the United States will be the last one for RMB internationalization.

The CEO's commendation for community service of the Lukfook Group said the internationalization of RMB is progressing steadily, but there are yet plenty of risks in relation to the volatility of the market. The financial turmoil in Southeast Asia in the last decade is a good lesson to be learnt. Moreover, it is believed that with the trend of further liberalization of the RMB exchange rate and interest rate, as well as proper risk management, the RMB currency will be continuously opened up internationally.

3.8 The development of the RMB offshore centre in Hong Kong

Vice President of the BOC (Bank of China) Hong Kong said as the largest RMB offshore market with about 70 percent of the offshore RMB trading in the world, Hong Kong's daily trade of RMB is approaching USD 35 billion. This better reflects the supply and demand of RMB in the offshore market.

Gold has the dual functions of investment and industrial usage. The RMB denoted gold will further enhance its functions in the internationalization of RMB gold.

People's Bank of China defines the middle exchange rate for RMB through the China Foreign Exchange Trading Centre every day, and the daily fluctuation of the exchange rate is limited within 2 percent. At the same time, People's Bank of China makes reference to the RMB exchange rate in the Hong Kong offshore centre to test the supply and demand condition of RMB, the change in RMB exchange rate due to economic data and the speculative trading of RMB.

3.9 The effect of RMB gold arbitrage on RMB exchange rate

The Chief Dealer in Forex & Bullion Dealing of HSBC said that the capital market in Mainland China is not yet completely open and the RMB cannot be exchanged with foreign currencies without any control within the country. As a result, there is a need for offshore RMB markets. In an offshore market, RMB can be freely exchanged with other currencies, and the respective exchange rates can provide a good reference for the market value of RMB. At present, plenty of Chinese and international financial institutions have established their branches in Hong Kong and Shanghai, and this can facilitate arbitrage in RMB in both onshore and offshore RMB markets. This has further reduced the differences between RMB exchange rates in onshore and offshore markets.

At the end of 2012, the market was pessimistic towards the future of the newly developed markets, and the exchange rates of these newly developed markets, including China, dropped in comparison to the USD. The onshore RMB exchange rate had also dropped beyond the fluctuation threshold continuously for more than ten days, and the exchange rate against the USD was around RMB 6.3. During that period, the offshore RMB exchange rate was not far off. Finally, as there was no significant profit that could be obtained through the short-selling of RMB, the market speculation activities halted and the RMB exchange rate resumed back to its normal trend.

With further relaxation of limits of fluctuations of the RMB exchange rate, there will be price differences between RMB gold and USD denoted gold owing to various reasons, and therefore there will be hedging and arbitrage trading focusing on this price difference. In the market, hedging and arbitrage trading are everywhere when there is a price difference for the same product. Here, the RMB gold arbitrage in the RMB offshore centre will have two major functions.

First, it can provide the real value of RMB gold. Second, it can provide the real value of RMB.

The Permanent Honorary President of the Chinese Gold & Silver Exchange Society anticipates that in the near future there will be arbitrage trading between onshore RMB gold and Hong Kong offshore RMB gold. Owing to the supply and demand of gold in the onshore gold market, there will be a premium or a depletion of the respective gold prices against that in the international market. As the RMB gold products offered by HKGSE have the same composition as gold products offered in Shanghai, this enhances the convenience for arbitrage and hedging of gold prices between these two locations. These arbitrage trades in turn will further improve the transparency of the RMB gold pricing.

Furthermore, during the process of RMB internationalization, the interest rate of RMB also plays an important role. In the RMB offshore centre, the lending market of RMB gold will also provide an important reference to the RMB interest rate.

4 Gold market and impact factors analysis

Gold is a noble among metals and a favorite in the economic world. It plays an important role in currency system evolution. Gold is considered a reflection of the ultimate non-governmental wealth. Gold has been a core component of the currency system because of its unique physical and chemical properties, but since the development of modern economy and the financial industry, gold's core position in currencies has been dented. In the 1970s, the International Monetary Fund (IMF) canceled gold's currency function in legal form. After that, gold became one of the most important investment avenues in the financial field due to its special exclusive value. Although gold is not used anymore in trade settlements, it is an acceptable settlement means for both parties in final balancing of payments. This chapter tries to analyze the factors that influence gold price and the price mechanism of gold so as to provide a basis for the judgment of gold price trend in the future and a reference basis for gold investment in theory and practice.

4.1 Multiple attributes and pricing mechanism of gold

4.1.1 Commodity attribute of gold

Gold is one of the rarest and most precious metals. In ancient times, gold was used as the main raw material for jewelry, utensils and apparatus. Gold has always represented social status and wealth. Gold texture is soft, so it is easily abraded and turned into super fine powder. Its density is 19.32 g/cm3, which is very high. Because of the feature of good extension and pressure, gold is easily forged and extended. Gold has strong resistance to corrosion and oxidation with stable chemical structure. Gold plays an important role in traditional industries and modern high-tech industry due to its good physical and chemical attributes. In medicine, gold alloy can be used to produce false teeth, tooth sockets and various gold salt preparations to cure skin diseases and rheumatic arthritis, etc. In the aerospace industry, gold and its alloys are applied in the driving mechanism of optical instruments, temperature control mechanisms, satellite-rocket separation mechanisms and mechanical system of atomic energy, etc. In the electronic industry, gold and its alloys are widely used in electric contact material, resistance material, welding material and temperature measurement material. In the chemical industry, gold can be used as a catalyst to accelerate reactions and as an additive to improve the texture of chemicals. With technical progress, gold has even wider applications prospects.

4.1.2   Currency attribute of gold

Gold is endowed with a social attribute – a currency in circulation, due to its unique natural attribute. As a currency, gold has a long history (see Table 4.1). Gold could be used for settlement of international trade and as a national reserve. During the period of the gold currency standard system, gold acted as a strong currency attribute. During the period of the Bretton Woods system, the USD was linked up with gold. 1 ounce of gold could be exchanged for USD 35. The USD was dominant in the international currency system. The United States must have sufficient gold as reserve when distributing the USD. At the time of the Bretton Woods system, gold, as a reserve asset for various countries, had a fixed official price. After that, demonetization reform was conducted due to disintegration of the Bretton Woods system. After that, gold still had the currency function, as it occupied a huge proportion in currency reserves of various countries and the IMF and was also an international settlement currency. Although gold seldom circulates in settlement nowadays, it is still accepted by various countries as the final guarder: the final means of payment in case of emergency requirements. Therefore, gold can be considered a quasi-currency nowadays.

Table 4.1 Currency gold for main countries from 1850–1930

Year	1845	1850	1855	1860	1865	1870	1875	1880	1885
Central Banks/ Treasuries Stocks (ton)	84	109	188	185	347	713	1089	1151	1536
Gold Coins in Circulation or with Commercial Banks (ton)		1300	1350	1725	2300	2650	2835	2976	3414	3439

Year	1890	1895	1900	1905	1910	1913	1915	1920	1925	1930
Central Banks/ Treasuries Stocks (ton)	3368	3350	4090	4710	5909	8098	9356	11295	13892	16469
Gold Coins in Circulation or with Commercial Banks (ton)	3439	3368	3350	3916	4699	3383	3298	2805	1565	984

Source: Timothy Green. Central Bank Gold Reserves, Centre for Public Policy Study, World Gold Council, 2003.

4.1.3 Investment attribute of gold

The financial attribute of gold is the extension and development of its commodity and currency attributes, which is the reflection of gold's charm in the new period. After the disintegration of the Bretton Woods system, gold's demonetization provided gold and its derivative product markets a huge development space. As a good financial investment asset in the new period, gold is favored extensively by investors.

As a peculiar investment value of financial assets, gold is reflected in the following three aspects:

First, like other staple commodities, gold features good anti-inflation and value preservation property. Quite a few researchers have indicated that there is a positive correlation between gold price and inflation. Gold price has been a leading indicator for inflation. Some articles have also pointed out that gold price has a smaller rising range than other staple commodities at the beginning of inflation. In later periods of inflation, the rising range of gold price was greater than other commodities and far more than the inflation rate (Akgiray et al., 1991). The possible reason for this might lie in that inflation is usually the reflection of rapid economic growth. At the beginning of inflation, prices of staple commodities such as petroleum rise significantly due to the influence of radically increasing commodity demand. The commodity demand of gold is relatively stable, and the gold price rises in a limited way. After inflation, it usually indicates economic downturn and recession. The prices of staple commodities such as petroleum decline significantly due to the impact of decrease of commodity demand. At this moment, gold begins to perform its hedge property and shows a pattern different from prices of other staple commodities. Therefore, gold has stronger value preservation and increasing property when compared with the general staple commodities.

Second, it is contrary to the price trend of main financial assets such as stock, which is another important characteristic of gold. Different from the fruits generated from ordinary financial assets, gold does not produce any fruits. On the contrary, some storage cost needs to be paid. During normal economic periods, stock produces a dividend, bond has some nominal interest rate, and the opportunity cost of gold investment is relatively high. Therefore, more people invest in other financial assets such as stock, which makes other financial markets flourish and results in the decline of gold price. During a crisis, the traditional financial assets markets such as stock, bond and foreign exchange are relatively sluggish, and the opportunity cost of gold investment is relatively low. Gold price is expected to rise under the pressure of inflation. More people tend to invest in gold, which results in the downturn of other markets and the radical increase of gold price. Based on this characteristic of gold, it is also a good asset that improves investment portfolios and adjusts risk structure. It can improve yield rate and stability of the whole asset portfolio.

Finally, gold offers strong hedging and is a reliable and stable investment variety. On the one hand, gold is a rare precious metal and has its own value with a limited price decline. On the other hand, gold is an ultimate currency accepted throughout the world. It has strong liquidity and is the exclusive asset that is not realized through state credit or company promise. Therefore, it is a good hedging asset. In recent years, with the implementation of a flexible exchange rate system and the advancement of globalization of international economy and trade, financial bubbles and economic instability are witnessed frequently even as the financial industry has developed well. People need reliable and stable investment varieties like gold.

4.2 Pricing mechanism of gold

Gold consumption and the investment market have developed rapidly. The international financial investment market makes gold price fluctuate in a moment and period manner, and then the demand for research and analysis of the fluctuation mechanism of gold price appears. Meanwhile, gold has a social attribute with the connotation of civilized history and the characteristics of a financial investment tool. Gold price is affected not only by the change of supply and demand balance but it also contains the original mechanism of the civilization's history and price mechanism of financial investment, which influence and restrict the evolution of direction of gold price at given moments or periods of time.

The gold market is a global one in which the price is transparent. Gold prices in several global gold bullion markets are referenced and influence each other, which renders a strong capacity of balancing inter-market price differences. Basically speaking, gold price is determined by supply-demand correlation and currency supply. Since the delinking of gold from the currency system, the normal gold price mechanism should be defined in the framework of neoclassical economics in which price, cost, benefit, supply and demand are mutually balanced. Due to its commodity, currency and investment attributes, gold is different from other metals, which makes the four factors (international politics, international economy, international finance and international investment) become the most important influencing forces (the four influencing factors for gold price originate from gold's status and influence in the international monetary system as well as its investment attribute). Only when gold shows its commodity attribute, this normal price mechanism comes into play.

Gold price is influenced by its currency attribute in international geopolitical situations. However, in the era of credit currency, currency is only a symbol relying on the apparent guarantee of the government; it is essentially the liabilities of issuers to the holders. When the international political situation becomes unstable, currency holders begin to doubt the government's liquidity. At this moment, gold price marked by paper currency gets adjusted, and other factors are weakened as well.

International economic situation refers to macroeconomic development of important economic entities and includes economic growth, supply and demand of important economic resources, social employment and price coefficient in relevant countries. Production, circulation and configuration of important economic resources in contemporary society have been globalized. One of the important functions of a country is to ensure the safe supply of important economic resources required for social and economic operations. When there is a conflict between supply and demand of important economic resources throughout the world, it results in strategic adjustment of resources in important countries and important resources gathering regions. This kind of adjustment also influences the stability anticipation of these countries and thus causes the relative price adjustment between gold and (paper) currency.

International financial situation refers to the status of financial development of international important economic entities and includes a series of development trends of financial markets and observable financial variables such as currency supply, distribution structure of financial assets, capital market, currency exchange rate, interest rate and investment market. There are two different mechanisms for transmitting its influence to gold price: first, when political, economic and financial factors trigger financial crises in important countries, gold shows its currency attribute which imposes mid- and long-term influence on gold price. Second, when the adjustment of important capital markets of economically advanced countries results in structural adjustment of unemployed capital, it imposes a short-term influence on gold price.

Structural adjustment of unemployed capital in the international investment market refers to large-scale adjustment of currency, capital and futures distributed in the international financial investment market as well as stocks and flows of distribution structure of unemployed capital among gold and financial derivatives markets. When international unemployed capital enters or exits from gold and financial derivatives markets due to profit pursuit, different fluctuations occur in exchange rate, capital market trend, futures and gold price and deliver to gold market to form the fourth-level factor influencing the price trend. This factor is dominant only when the above three factors are relatively stable.

Although the global gold reserve that typically represents the currency attribute of gold only occupies 18 percent, social gold reserves occupy 16 percent and the commodity gold occupies above 60 percent, the commodity attribute of gold is still controlled by currency and investment attributes at a micro level, reflecting internal political, economic, financial and investment situations at super macro and macro levels.

4.3   Analysis of factors influencing gold price

4.3.1   *Influence of supply–demand relation on gold price*

4.3.1.1   *Gold supply*

Gold supply is mainly divided into mined gold, recycled gold and central bank gold (sales). Gold supply is mainly from the production of mined gold. That is, gold is supplied from production as a general commodity. The ascertained global gold resource totals 89,000 tonnes, and its reserve base is 77,000 tonnes. The gold ores are distributed unevenly in the world. 80 percent of mineral resources are mainly from South Africa, the United States, Canada, Russia and Australia. The gold mineral resource is most abundant in South Africa, which has 40 percent of the mineral gold in the world. It takes 5–7 years from finding mineral gold to mine it. Therefore, gold price imposes little influence on the supply of mineral gold, and the supply of mineral gold is reasonably elastic but with a tremendous lag.

Supply of recycled gold occupies one-quarter of total international gold supply. Recycled gold refers to recycling and reproducing waste gold to be supplied in the gold market. Its main sources are recycled products such as scrap jewelry. Compared with central bank gold sales and gold mining, the supply of recycled gold is relatively sensitive to the change of gold price. When the gold price rises, suppliers increase gold supply in order to make a profit. Otherwise they reduce the supply by holding on or the general public is simply disinclined to sell pieces of jewelry against cash. Therefore, the supply of recycled gold is more elastic. The anticipation of gold price is the greatest factor that influences the supply of recycled gold. When the suppliers of recycled gold anticipate that the price will rise, they wait for the best time to sell it. When the suppliers of recycled gold anticipate that the price will decline, they sell it in large amounts. Change of economic prosperity is another factor to influence the supply of recycled gold. During economic recession, people do not choose financial commodities that entail high risk, such as stock and funds, but invest in gold due to the smaller risk. During economic prosperity, people are in no hurry to sell gold, supply of recycled gold decreases and price falls. The supply of recyclable gold shares positive correlation with rise and fall of gold price. After 2007, the supply of recycled gold increased considerably and unexpectedly, and it slowed down in 2009. The supply of recycled gold in 2009 increased by 62 percent compared with 2007. People expected an excessively large decline in gold price and in anticipation they were in a hurry to sell gold. From 2009 to 2011, many economic reports showed that the downturn in global prosperity resulted in suppliers of recycled gold changing their expectations of gold price, and the supply of recycled gold remained about 1,700 tonnes. When gold price moves upward, the supply of recycled gold increases: the two have shared a positive relationship.

Up to the end of 2012, gold reserves of central banks of various countries and multinational organizations (e.g. IMF) totaled 31,491.1 tonnes, accounting for about 19 percent of global gold inventory. After the disintegration of the Bretton Woods system, the total gold reserves of national governments showed a declining trend. Central Banks' gold holding used to be a major source for gold supply. At the end of 1990s, quite a few European countries sold significant amounts of gold reserves which resulted in gold price falling to the lowest level. To prevent gold price from being crushed by Central Banks, eleven European countries and the European Central Bank signed the Central Bank Gold Agreement on 27 September 1999. It was stipulated in this agreement that the signing countries could undersell only 400 tonnes of gold every year in five years. After that, the Central Bank Gold Agreement (CBGA) was extended from the first phase to the third phase. The first few years when the CBGA was executed, gold selling quota was just used up. From 2008, this situation changed to some extent. In 2008, central banks sold a total of 279 tonnes of gold. In 2011, there were no gold sales by central banks, and 457 tonnes of gold were purchased for reserves. It can be seen that the rise of gold price has aroused the concerns of various governments for gold value since 2008.

Historically, gold supply has always been limited by mining output and existing inventory, and the supply is stable. Data show that the gold supply after 2000 has also been basically stable. The annual gold supply remains about 4,000 tonnes. The supply of mineral gold fluctuates minimally, remaining between 2,600~2,900 tonnes every year. This is mainly attributed to limited minerals and restraints of mining technology. Moreover, the supply of recycled gold fluctuates significantly. This is mainly attributed to the great influence of gold price on the supply of recycled gold. In the bull market for gold after 2000, the supply of recycled gold kept increasing from 749 tonnes in 2001 to 1,694.7 tonnes in 2009. The supply from central banks remained above 360 tonnes from 2001 to 2007. After 2008, central banks began to purchase a certain amount of gold due to the rising price, concern about the USD or concern over US abuse of its power. Also, central banks became more risk averse and this encouraged a little more buying but especially stopped selling from Europe – the latter also dried up because the sellers had started to look very silly in the light of continuous gold price increases well above their realized prices! It can be seen that the gold supply is relatively stable.

4.3.1.2 Demand of gold

Factors that impact demand for gold are mainly jewelry, industry and investment.

The demand for jewelry accounts for about 50 percent of total market demand. In particular, the main demand for gold jewelry is from Asian countries that use gold for weddings. For example, the consuming countries for traditional gold jewelry include India, Saudi Arabia, China and Turkey. As a populous country, India's jewelry industry imports 15–20 tonnes of gold every month. Gold jewelry is closely related to the income level of consumers, and it is an elastic commodity. As the consumers' incomes increase, gold demand increases as well. As commodity prices fall, people's purchasing power increases and this drives the increase of gold jewelry demand. The demand of ornamental gold imposes great influence on gold price, showing periodic and seasonal influences. Every year, gold demand increases during religious festivals in India, Lunar New Year in China, and Christmas in the West. In particular, demand for ornamental gold in the first and fourth quarters is the highest. Until recently, demand for gold jewelry had showed a declining trend because of the rising price of gold. Moreover, the global economic situation was bad, which had resulted in the reduction of people's incomes.

Industrial demand accounts for about 10 percent of total market demand. Due to its good physical attributes, gold is widely used in industry, particularly high-tech industry. In industrial fields, the electronic industry and dentists are the biggest users of gold, taking up 71 percent and 9 percent respectively. Other industries account for the remaining 20 percent. In the recent 10 years, the total industrial demand for gold has remained at about 10 percent of total gold demand, which is relatively stable. In 2008 and 2009, gold used in industry decreased by 6.72 percent and 4.65 percent, respectively, after an increase the two years prior. Due to the influence from economic recession and technology changing, overall reduction of industrial output should be the major reason. With economic development and progress of social sciences, precious metals with low price and unstable performance are expected to be replaced with gold with high price and stable performance when raw materials are selected for electronic products. The increase of industrial demand for gold from the consumers will promote the rise of gold price. These two have a positive correlation.

Gold is a commodity with an investment property. The anticipation of a rise in gold price in markets also results in speculative demand. Gold investment can be divided into entity investment and financial commodity investment, such as gold futures Exchange Traded Fund (ETF). Gold ETF and gold price move in sync with each other, which obviates the need for storage and thus improves the demand for gold. Gold ETF provides investors with a convenient transaction channel. Since its sales in 2004, gold ETF has attracted hundreds of billions of dollars to flow into the gold market, which is the most direct factor responsible for the rise of gold price in recent years. Gold is also a commodity with a value preservation property. During economic downturns, value preservation function of gold becomes more significant. People do not invest in currencies but purchase gold with value preservation functions. After 2007, gold bar investment has increased gradually. Since the financial crisis had just broken out, investors increased gold bar investment, which carried low risk. Moreover, the anticipation of the rise of gold price was gradually strengthened, resulting in increased investment demand, and the gold price rose.

4.3.1.3 Analysis of gold supply and demand

The total gold supply does not change too much every year. Moreover, demand and supply change in the same direction every year. Therefore, it is believed that gold supply in the mid-term influences gold price but imposes little influence in the short term. It is not the main factor of influence for the short-term fluctuation of gold price.

The difference between supply of and demand for gold was the largest in 2009, but the change of gold price did not slow down. Therefore, the supply surplus cannot be analyzed in isolation. Price can go up even with a larger surplus to be absorbed if investors or central banks are keener to buy and hold the metal. Conversely a smaller surplus of bullion in the market could be consistent with lower gold prices if buyers of bullion are not interested in gold and the price needs to fall to a level where "value investors" are motivated to come in and buy.

In the long run, a rise in gold price makes producers increase production, but for the short term it is less clear cut. For example, small and especially artisanal producers who are more nimble will "make hay whilst the sun shines". Larger producers cannot do this and may even seek to extend their mine lives by mining less or at least lowering the average grade they mine.

Jewelry is the largest component of demand for gold. Due to the influence of consumers' habits, it is less sensitive to price. This indicates that with the rise of gold price, people's demand for jewelry decreases and vice versa. In recent years, the demand for gold jewelry has declined due to such factors as high gold price. Jewelry accounted for 71 percent of total gold demand in 2002 while it was only 45 percent in 2011. Industrial demand and gold price have a highly positive relationship, and the correlation coefficient is 0.929. This indicates that scientific progress enables multiple products to use gold with stable performance. The industrial demand for gold is increasing, resulting in the rise of gold price. However, it accounts for a small percentage of the total demand, so its influence on total gold demand is limited. Investment demand and gold price show a highly positive relationship, with a correlation coefficient of 0.956. Due to the poor global economic situation, gold price keeps rising. To reduce the risks, investors add gold to investment portfolios, anticipating that the gold price will rise. Therefore, its proportion in total investment demand is increasing year by year.

The significant rise of gold price in recent years indicates that gold supply and demand are only a basis for analysis of gold price and cannot fully explain the reason why gold price fluctuates. If important events such as economic instability or wars occur, people rush to purchase gold and then supply and demand becomes a major influencing factor for gold price. As a special commodity with many attributes, some other global factors also need to be taken into consideration, besides the price. For example, the USD index, oil price, stock price coefficient, federal fund rate, etc., influence the changes of gold price. The following is an analysis of long-term balance and influence of these variables on the fluctuation of gold price.

4.3.2 Macroeconomic factors

4.3.2.1 Economic growth

Generally, GDP is used to measure economic growth. When Gross National Product (GNP) rises, it means that the economy is in a growth phase. Hot money from around the world floods into the country to seek higher investment returns. At this moment, value of the currency of the country rises. Meanwhile, gold demand for industrial development of the country increases and there is private hoarding. Part of this demand gets reflected in the balance sheet of supply and demand. They are visible and can be easily measured and integrated into the gold market price. However, other private gold storages are conducted gradually as the national income per capita increases. They are invisible and cannot be easily detected and reflected in market price. As time goes by, gold price faces a short squeeze. Otherwise, if GNP falls, economic growth slows or enters into a declining stage. The capital of the country might flow out, and its currency value decreases. The gold demand in industry decreases, and people's capacity of purchasing gold gets weakened.

When the GDP of a country grows too fast, its government might raise the interest rates to tighten the currency supply in order to ensure the economy doesn't get too overheated. At such moments, the currency strengthens. People are more willing to hold the currency instead of gold. In case of economic recession, the government might decrease the interest rate to stimulate the economy. Accordingly, the currency supply increases and inflation rises, and the attraction of the currency weakens. People will be willing to hold gold to pursue assets value preservation and appreciation. In the long run, the residents all over the world treat gold as a quasi-currency. Normally, economic growth is accompanied with inflation. To maintain its purchasing power, residents purchase gold to avoid asset depreciation. Therefore, whether the currency of a country is strong influences the conversion between currency and gold held by people. In addition, the "wealth effect" for jewelry demand, especially in developing Asia, is very important. For example 24 or 22 karat jewelry is purchased in large measure as an investment.

4.3.2.2   Inflation

Inflation means currency in circulation exceeds demand, which results in currency depreciation and continual and widespread rise of commodity prices. The influence of inflation on gold price is specifically analyzed from long-term and short-term perspectives.

From the long-term perspective, although inflation is always a management highlight of various national governments, it is commonly believed in academic circles that creeping inflation is not harmful but beneficial to promoting economic development in case of insufficient effective demand. Therefore, governments can usually accept its long-term existence. In fact, post-Bretton Woods currency, i.e. paper currency, always keeps depreciating slowly. After demonetization of gold, its price is marked with a general commodity. Continual and slow depreciation of currency makes nominal prices of commodities, including gold, increase steadily in the long run.

From the short-term perspective, the influence of inflation on gold price is determined according to the intensity of inflation. Both do not necessarily show an obvious and positive correlation. When there is galloping or virulent inflation, markets become apprehensive of scarcity. Due to its unique value preservation and hedging attributes, gold is favored by people, which makes gold price rise radically in the short term. Let's take the global inflation at the end of 1970s as an example. At the end of 1978, when the second oil crisis broke out, oil price rose radically, and it was a severe inflationary period throughout the world. In the same period, gold price rose from USD 207.7/oz at the end of 1978 to USD 850/oz on 21 January 1980. However, when the inflation rate is in the normal mode and what is predicted by people, then the influence of inflation on gold price is relatively weak. For example, after President Regan came into power in the beginning of 1980s, the US inflation rate was well controlled within 7 percent. Double-digit inflation in 1970s was not seen anymore. The correlation between inflation rate and gold price was no longer obvious.

4.3.2.3 Interest rate

Interest rate represents the ratio between interest and principal at a certain period and is usually expressed in percentage. Interest rate is usually regulated by central banks, and the USD is managed by the Federal Reserve Board. Interest rate is considered one of the powerful currency policies of a country. When the economy is overheated and inflation rises, central banks increase the interest rate and consumers select savings with higher interest rates for future consumption. When an overheated economy and inflation are controlled, consumers reduce savings and increase their current consumption. Therefore, interest rate is an important and basic factor in economy. Interest rate is divided into nominal interest rate and real interest rate. Nominal interest rate refers to the interest rate issued by the monetary authority of a country. In the economic operation, what really functions is real interest rate (effective interest rate). Real interest rate refers to the interest rate after providing for inflation. It can be obtained by subtracting the rate of inflation from the nominal interest rate. For example, a benchmark nominal interest rate is the Fed Fund Rate from the United States Federal Reserve Board. Compared with other investment tools, people hold gold for value preservation and do not conduct short-term investment. They do not make a profit by gold price differences at different time points. As a whole, gold is unrequited. Under this circumstance, interest becomes the opportunity cost of holding gold. However, purchase of gold has to be abandoned when the real interest rate is higher since the opportunity cost will be higher. When the real interest rate is lower, the opportunity cost is lower. Thus the real interest rate and gold price have reverse trends.

From another perspective, interest rate has an important influence on currency supply. The increase of interest rate makes the currency multiplier reduce, which reduces the currency supply. Then, the purchasing power of the USD is improved, and gold price in USD reduces accordingly and vice versa. The imposing path for the influence of interest rate on gold price is relatively long, and its influence is not direct. Meanwhile, the relationship between interest rate and inflation is so complicated that evidence can hardly be found to support this principle. During the period of overheated economy and inflation, a country usually increases the interest rate. In the background of common depreciation of commodities, the function of gold as a value preservation means is prominent, gold demand increases, and its price rises. The reverse trend of interest rate and gold price appears only after the inflation is restricted.

It is found that before 2001, the trend between gold price and real interest rate in the United States was basically consistent, showing a positive correlation. After 2001, the trend changed radically. The real interest rate fluctuated and declined, on the whole. However, gold price continued to rise and hit historical highs. In the most recent ten years, the relationship between gold price and real interest rate has not been so clear and stable. Their correlation coefficient is -0.29. Comparatively speaking, this causal relationship coefficient is at a relatively low level.

4.3.2.4   Important influence of the US dollar

After the disintegration of the Bretton Woods system, the international positioning of the USD remained the same. In foreign exchange reserves of various countries, the USD still occupies a very important position. In international trade, the USD is still the main charging unit and transaction intermediary. Some staple commodities such as petroleum and corn are priced in USD. Currently, the United States is the most economically developed country in the world, and the USD is guaranteed by US national credit. Therefore, the long-term price of gold is determined by the specific value of USD. We can abstract gold to be the currency of a virtual country. The USD is the currency of the United States, and the specific value is the exchange rate between the virtual country's currency and the USD. The USD's value is protected by US laws but it is still a paper currency, and its value is often influenced and weakened by US economic recession and extra paper currency printed by the Federal Reserve Board. The internal value of gold is stable, so gold price is mainly determined by the currency value of the USD. When other countries are not confident about the USD, gold price will tend to rise. When they are strongly confident about the USD, gold price will not rise or even will fall.

The confidence in the currency of a country is based on strong national power, stable currency value, sound financial market and the inertia factor. Strong national power is a comprehensive concept that is closely related to technical strength, population and geography. It includes such factors as technology, culture, education and resources. The international positioning of the USD is inseparable from America's economic strength. The United States is a great resource and agricultural country and takes the lead in technological and high-tech industries in the world. We can select the US GDP index to judge the comprehensive strength of the United States. Stable currency value refers to stability of purchasing power of the USD, including internal purchasing power and external purchasing power. Stability of internal purchasing power of the USD means that the US inflation rate remains at a relatively low level, and the residents' capacity for purchasing domestic commodities and services remains stable after holding the USD for a period of time. External purchasing power of the USD means that the US exchange rate remains stable. When the residents of other countries hold the USD for a period of time, their capacity for purchasing commodities and services remains stable. We can select USD index, federal fund rate and the US inflation rate to judge the stability of the USD. The United States has the most developed financial market in the world, but the US financial market was rather unstable during the sub-prime crisis. We can select the Dow-Jones Average as a test index for the US financial market. The USD has been an international currency for a long time. No other currency has been able to replace the USD over a long period. Therefore, inertia factor is not taken into consideration.

4.3.2.5 Mechanism of the USD exchange rate influencing gold price

It is commonly believed in extant research that the USD exchange rate is one of the important factors influencing gold price. Review of the transition history of international gold price shows that the USD exchange rate and international gold price show a negative correlation most of the time. As the first reserve currency in the world, appreciation and depreciation of the USD exerts significant influence on the US economy, world economy, global commodities and capital markets. At the same time, gold is a special substance that integrates commodity, currency and financial attributes. Under this circumstance, the function of the USD exchange rate for gold price is complicated. In order to clarify the influence of the USD exchange rate on gold price, this chapter analyzes the mechanisms of influence according to different attributes of gold.

INFLUENCE OF USD EXCHANGE RATE ON GOLD PRICE UNDER THE
COMMODITY ATTRIBUTE OF GOLD

Under its commodity attribute, gold is a general commodity. The influence of the USD exchange rate on gold price mainly lies in actual supply and demand of gold. Since international gold is denominated in USD, depreciation (appreciation) of USD directly results in rise (fall) of international gold price. When other relevant factors are not taken into consideration, gold price rises (falls) as USD depreciates (appreciates). According to the classical purchasing power parity theory, USD depreciation means the decline of purchasing power of USD. The nominal price of gold in USD will be lower than the actual value, which stimulates gold demand and drives the price up till the nominal price equals the actual value. On the contrary, appreciation of USD indicates that the nominal price of gold in USD is higher than the actual value, and gold demand decreases, which makes the gold price fall.

INFLUENCE OF USD EXCHANGE RATE ON GOLD PRICE UNDER
THE CURRENCY ATTRIBUTE

After the disintegration of the Bretton Woods system, gold exited the circulation field and lost its function as circulating currency. Until now, gold still has currency functions such as storage means and payment means, and it is the fifth hard currency in the world, besides the USD, EUR, JPY and GBP. The mechanism of the influence of the USD exchange rate on gold price under the currency attribute of gold lies in the substitution effect between them.

According to the management theory of international reserves structure, reserve assets need to realize an organic unity combining security, liquidity and profitability. Because of the currency backed by gold having been replaced by paper currency, USD and gold are now independent of each other. First, in terms of security, to hold any currencies is actually to have creditors' rights in the currency issuing country, and its value is influenced by economic and political factors in the debtor country. Gold is a carrier of commodity value, and it is not influenced by any economic and political risks of a country. Specifically, strong USD means America's economic prosperity. The US government's guarantee for its debts is viewed as strong, which weakens gold's value preservation and the performance of its hedging attribute. On the contrary, USD weakening means economic downturn in the US, implying a weak guarantee capacity for its debts. As a reserve asset, the security of gold can be thus highlighted. Second, in terms of liquidity, credit currencies such as USD are still the main means of circulation during periods of stable international economy and politics, and they are of strong liquidity. External payment by gold needs to take gold market as a medium, and its inconvenience has weakened its liquidity. During unstable periods of international economy and politics, credit currencies fully depreciate when virulent inflation occurs. Then, credit currencies such as the USD weaken greatly. Gold is widely accepted since there is no depreciation risk, and its liquidity is strengthened accordingly. Finally, in terms of profitability, the difference between gold and credit currencies including USD lies in that interest cannot be earned by holding gold, and capital premium can only be obtained from the rise of gold price. Therefore, credit currencies including USD offer good profitability. The market price of gold is sluggish and its profitability is poor due to restraints of currency and financial attributes. During a crisis, various countries cut interest rates to tackle the crises and adopt easy currency policies. The yield rate of credit currencies is relatively low. During high inflation, economic downturn and risk intensification, gold price is expected to rise. At this moment, to store gold will gain high profitability.

To sum up, strong USD means strong security, liquidity and profitability, which weakens gold's value preservation and hedging attributes. Due to poor liquidity and profitability, various national governments tend to hold USD and sell gold, which results in the decline of gold price. On the contrary, the continual weakening of the USD means reduced security and profitability, and gold security comes to the fore. In addition, gold profitability from bear market to bull market rises significantly, which further drives various national governments to purchase gold to replace USD assets and makes the gold price rise. In stark contrast, as "strong USD" policy was replaced by "weak USD" policy in the beginning of the twenty-first century, various national governments undersold gold at a smaller scale. In 2008 and 2009, central banks' gold sales totaled 279 tonnes and 34 tonnes respectively. In 2011, 457 tonnes of gold was purchased.

However, it is pointed out that the USD is still the first international reserve currency. There are no other credit currencies that can contend against USD. During global crises, the USD is a relatively safe hedging asset. To prevent the crisis from spreading and causing thorough breakdown of the currency system, various national governments have to keep holding USD assets, which results in the abnormal, continual and simultaneous rise of the USD exchange rate and gold price since the global financial crisis in 2008.

The financial attribute of gold refers to a financial asset which is favored extensively by investors because it facilitates value preservation and hedging against adverse movements of main financial assets. From this point and under the financial attribute of gold, the mechanism of the influence of the USD exchange rate on gold price mainly contains the following three aspects:

> First, the absolute depreciation effect of paper currency. At present, various countries use credit currencies, so global inflation is unavoidable. Since the twenty-first century, the United States has adopted a weak dollar policy and has issued too many dollars, which unavoidably results in expectation of weakness in USD in the long term, because of higher US inflation. On one hand, it brings global inflation and reduces currency value. Prices of commodities including gold rise directly. On the other hand, the persistent expectation of long-term decline in currency value stimulates the financial attribute of staple commodities (such as gold) that can resist inflation. In such circumstances investors step in and buy such commodities for value preservation and appreciation, thereby increasing demand, which further improves the price.

Second, the asset transformation effect: USD depreciation implies a slump in the US economy and that in turn means decelerated growth of the world economy. A worldwide economic slump usually indicates downturn of traditional capital markets including stock, bond and foreign exchanges. Since gold and most other financial assets generally move in opposite directions, a lot of capital flows into the thriving gold market, from traditional capital markets, for seeking investment opportunities, driving the gold price up in the process. Therefore, decline of other assets and rise of gold price is the outcome. This is also a cause for the expanded differences between the two. In addition, the scale of the gold market is relatively small when compared with traditional foreign exchange, bond and stock markets. The flooding of a lot of capital can easily result in a radical increase in gold price in the short term.

Third, the psychological anticipation effect. Keynes was the first to introduce the anticipation theory to the study of economic issues. After the rise of the rational anticipation school in the 1970s, anticipation analysis became the mainstream for economic analysis. It is believed in anticipation theory that people judge the future trend of economic variables according to the past level and direction of economic variables so as to adopt the corresponding actions to maximize their personal benefits. In the background of continual depreciation of the USD, all seem to be pessimistic. The depreciation of the USD results in anticipation of its continual depreciation, which generates expectations of the corresponding inflation. This psychological anticipation motivates investors to undersell the USD and purchase gold on a large scale. This accelerates the value of the USD in the short term, makes more capital flow into the gold market and drives the price to be out of control in the short term.

4.3.3   Other factors influencing gold price

4.3.3.1   An analysis of stock index impact on gold price

The Dow Jones Industrial Average is the most influential and widely used stock index in the world. It is compiled on the basis of some representative corporate stocks that are publicly traded on the New York Stock Exchange. Comprising an average of four indexes, it reflects the general trend of the US stock market.

The Dow Jones Industrial Average is negatively correlated to the value of gold. The correlation coefficient between these two is -0.28, based on tests by EVIEWS software during the observation period. The negative correlation can also be verified through the respective performances of the two markets at the early stage of the financial crisis in 2008. From early October 2007, when the financial crisis occurred, the Dow Jones Average went down from its peak of 14,000 to 8,450 in October 2008, a decline of 40 percent in a year. Instead of declining, gold price was up 12 percent in the same year, from USD 740/oz to USD 830/oz, with a record high exceeding USD 1,000/oz, which manifest the hedge function of gold.

The negative correlation between the Dow and gold price is mainly caused by three factors. First, as investment tools, stock and gold can be mutually replaceable to some extent. When stocks rise, the opportunity cost of investing in gold is high due to the falling gold price, caused by outflows from the gold market. Second, when the US macroeconomy remains upbeat, listed companies have healthy financials and huge corporate profits make the stock market rise further. Gold tends to perform worse during periods of fast macroeconomic growth. Third, when the US macroeconomy gains momentum, the relatively strong dollar index puts great pressure on gold price. Speculative funds are risky assets, so gold, as a safe haven asset, performs poorly.

4.3.3.2   Petroleum price

Petroleum is an important strategic resource for economic growth of industrialized countries as well as emerging market countries, and it constitutes an indicator to judge the world economic development. Today, as the blood of modern society, petroleum is known as "black gold". Petroleum production of OPEC countries accounts for two-fifths of the world's production and their exports account for three-fifths of the global trade which greatly influences the whole international petroleum market and petroleum price. Therefore, the petroleum price set by OPEC is usually regarded as standard for international petroleum. Both petroleum and gold play important roles and are priced in USD, the primary currency in international trade. So, when the USD changes, gold price and petroleum price also change in the same direction.

Historically, a rise in the price of petroleum is usually followed by an increase in global inflation. The rising price results in mushrooming of incomes of countries producing petroleum. That is to say, these countries trade petroleum for more USD. However, as the price in countries importing petroleum keeps rising, domestic spending by petroleum producers rises and prices in petroleum producing countries also go up. For the purpose of hedging and avoiding risks, some countries and investors increase the proportion of gold purchases as a hedging tool in periods of inflation. The rising demand for gold sends the gold price up. With the continued economic growth in America, India and Middle Eastern countries, major petroleum corporations have claimed that petroleum price is not likely to fall significantly in the near future, leading the international gold price to a trend of continuous rise, along with the fluctuation of petroleum price. The high petroleum price contributes to the increase of inflationary pressures and in the meantime, inflation reacts to petroleum price and drives the petroleum price up. As the best hedge during inflationary periods, gold is the hedge against inflation. So gold price goes up when inflation is relatively high. When the economic situation improves, petroleum price declines with the rise in interest rates. Meanwhile, the financial attribute of gold is weakened at different levels. The surge of gold price caused by the US financial crisis in 2008 and the European sovereign debt crisis that followed strongly illustrated the function of the financial attribute of gold.

In the past 30 years, average international price of gold was USD 300/oz and that of petroleum was about USD 20/barrel, which means 1 ounce of gold equaled 16 barrels of petroleum. To further explain the changing relationship between gold and petroleum price, this chapter analyzes the international gold price and petroleum price between 1973 and 2012.

The trends of international gold price and international petroleum price are synchronous. In the 1970s and 1980s USD depreciation and the petroleum crisis caused soaring gold and petroleum prices; from the 1980s to the beginning of the twenty-first century, petroleum and gold prices remained low for a long time. Since 2001, international political events such as "9/11" pushed the international petroleum price up almost linearly, and the inflation brought by the rising petroleum price sharply escalated the international gold price. Based on these historical stages, when petroleum price goes up, it's reasonable to expect bullish gold price. Petroleum price can be the signal of the rise or fall of gold price. However, the phenomenon that trends of petroleum and gold price move the same way cannot prove that petroleum price is a vital driving factor influencing gold price. There is no direct relevance between petroleum price and gold price, and the reason why they share the same long-term trend is that both are influenced by the same factors. These factors result in the rise and fall of petroleum price and gold price and maintain the positive correlation between gold price and petroleum price at 80 percent. However, this relationship is not absolute. For example, there were deviations found from 1975 to 1976, 1985 to 1986, 1998 to 1999, 2006 to 2007 and 2008 to 2009, which were caused by international political events.

4.3.3.3   Other major precious metals

According to the market mechanism emphasized by the theory of microeconomics, the law of supply and demand governs the long-run equilibrium price of commodities. But future pricing now dominates trading in major commodities. So the price of precious metals is inevitably influenced by how the futures markets are running and the influencing factors are developing in diverse ways.

The trends in the price of gold and precious metals are compared after observation of historical gold prices. At present, in the backdrop of interlinked global currencies, price of gold and silver are not affected by currencies because they are measured by purchasing power parity. Silver price didn't change much at the beginning of the century. By comparing the trends of gold and silver price since 2000, it can be seen that the change of gold price has been consistent with that of silver when specific circumstances are not considered (specific circumstances mainly refer to the abnormal fluctuation of silver price caused by manipulation of hot money).

As for platinum and palladium, different from gold and silver, they have never been used as currencies throughout history. Thus, when analyzing the price of platinum and palladium, policy measures such as intervention and monetization by central banks cannot be considered. So it is much easier to analyze them than gold and silver.

In the twenty-first century, platinum price has shown a tendency of increasing fluctuation just like its historical price. Different from palladium, platinum fluctuates without any impact on appreciation. The reason is that the underlying demand continues to push platinum price without structural changes influencing platinum consumption negatively. Global concerns on policies and the resultant regulations for safeguarding against pollution have made use of platinum as a catalyst a necessity. Therefore the influence of gold price on platinum is basically realized through economic growth. Besides, the consumption pattern of platinum is such that its price is more prone to supply crises. The liquidity of platinum is much less than that of gold and silver as participants in the industrial chain of platinum lack diversity and there are relatively few producers.

Moving trends of platinum, palladium and gold show that the relevance between them is not as direct as that between silver and gold. The degree of fitting of gold price and platinum price is 0.729964, and the degree of fitting of gold price and palladium price is merely 0.193467, which indicates a low degree of fitting between gold and palladium. It can be concluded from the analysis that the specific demand model of palladium decides the difference between the long term price of gold and platinum. From 2003 to late 2012 the trends and fluctuations of gold price and palladium price had some similarities, in spite of the little relevance between the two. That the reason might be selection of the sample cannot be ruled out. Generally, from 2000 to 2003, palladium fluctuated dramatically and its price was several times higher than gold. However, apparently palladium price had no impact on the steadily rising gold price and investors' focus on palladium was not influenced by gold price.

4.3.3.4 Emergencies (international crisis, war) influencing gold price

International political turbulence caused by wars and significant political events also explains the rising gold price. When a political situation is unstable, gold displays its "hard currency" attributes strongly and serves its hedge function, encouraging people to pile into the gold trade market to buy it in bulk, which results in a price surge. Second, when there is international political turbulence, relevant factors such as cost maintaining stability in the domestic economy during a war prompts investors to move into gold in a big way, increasing gold demand and price.

To further explain the influence from international political situations on gold price, this chapter selects the trend of international gold price from 1973 to 2010, analyzing the relationship between the two from two aspects, time and political factors.

Chronologically, since the Second World War, gold price has undergone two sharp rises and falls. It can be seen that the first rise and fall occurred between 1970 and 1980. The breakout and pacification of the fourth Middle East war resulted in the steep rise and fall in gold price. A key reason for the drop in gold prices at the end of the 1980s and more so in the 1990s was the collapse of the Soviet Union and the ending of the Cold War.

Considering the political factors influencing gold price, there are two main manifestations. One is paroxysmal situational changes that make gold price rise significantly on the day. For example, before 2 August 1990, the daily rise and fall in gold price was merely USD 1~3. But Iraq's invasion of Kuwait on that day pushed gold price up from USD 370.6/oz to USD 380.7/oz, an increase of 2.7 percent. The other is that the situation shows an obvious sign before change. Gold price rises gradually with people's expectations but starts to fall when the situation changes. This phenomenon is the so-called "War Premium". For example, in 2002, the Iraq crisis pushed gold price up to USD 382.1/oz before the war but price gradually fell after the war between the US and Iraq in 2003. The "War Premium" came into play before the war and the breakout actually put an end to the rising price.

4.3.3.5 Seasonal fluctuations of gold price

Gold price is also influenced by seasonal fluctuations. Generally speaking, gold price starts to fall in spring and remains sluggish in summer. In autumn spot gold is in great demand because of festivals around the world and it shows an upward trend throughout the second half of the year. Here are some examples. India is a major country for consumption of gold ornaments. Indian brides who mostly have weddings in spring or autumn have the convention that the dowry should include gold ornaments. As Christmas arrives, European countries experience great demand for gold ornaments. At the beginning of the year, when the Spring Festival is close, Chinese people need to purchase gold ornaments for relatives and friends, and their purchases push gold price up.

5 Internationalization of RMB and Hong Kong as an offshore market

At the end of 2008, the Chinese government formed a new international financial strategy comprising a "trinity" – promoting RMB internationalization, accelerating currency cooperation in East Asia and advocating the implementation of international currency system reform. Significant progress has been made in many aspects in terms of CNY internationalization. With acceleration of the CNY internationalization process, many problems have appeared gradually.

In such a background, the formation and development of the Offshore CNY Market in Hong Kong (CNH) is the starting point to effectively solve the problems related to CNY internationalization. CNH undoubtedly plays an important promotional role in CNY internationalization, establishing the investment and financing market of offshore CNY, forming of a stable, orderly and external CNY circulation system, perfecting backflow channels of offshore CNY and gradually realizing totally free convertibility of CNY. CNY internationalization requires a series of gradual measures. Offshore CNY market and cross-border CNY trade settlement are the main measures that are driving CNY internationalization at present.

Meanwhile, the introduction and development of CNH is conducive to enhancing China's status in the international financial field and strengthening China's discourse power and initiative for participating in international currency system reform. Moreover, it can gradually promote CNY use for international settlement and investment, enhance stability of CNY, realize exchange rate coordination and cooperation at higher levels in the region and boost the development of East Asian currency cooperation. In a word, the formation and development of CNH plays an important role in boosting China's international financial strategy.

5.1 International currency and currency internationalization

5.1.1 Definition, function and nature of currency

During the 1960s, scholars began to study the issues of international monetary and currency internationalization. With the decline of the US dollar hegemony, process of the yen's internationalization, the European monetary integration and the emergence of the euro as an international currency, currency internationalization issues have evoked the interest of more and more researchers. Currency internationalization and the conditions and paths of becoming an international currency are issues that have been examined in-depth from different perspectives.

A currency in the most specific use of the word refers to money in any form when in actual use or circulation, as a medium of exchange, especially paper money in circulation. Currency is a symbol of the sovereignty of a country and generally circulates within a country, but along with economic development and expansion of exchanges on a global scale, internationally acceptable currencies have risen. Scholars from various countries and international organizations have presented some academic points of view on the concept of international currency.

Cohen (1971) defined it from the perspective of monetary functions; the functions of the international currency are expansions of domestic currency functions in a foreign country. When the private sector and official institutions extend the use of one currency outside the country for a variety of purposes, the currency is on the path of development into an international currency.

Hartmann (1998) further extended Cohen's definition and classified different functions of the international currency. He believed that as a means of payment, an international currency acts as an intermediary currency within direct international trading and exchange of two other currencies as well as a foreign exchange market intervention instrument used by the official sector to balance international payments. As a unit of accounting, the international currency acts as a pricing tool for merchandise trade and financial transactions, and is used to determine the exchange rate parity by the official sector. As a store of value, the international currency is used by the private sector as the choice of financial assets, such as non-residents' holdings of bonds and deposits, while the official sector uses international currency denominated financial assets as reserve assets.

The International Monetary Fund (IMF, 1994) believes an international currency can function as a general measure of value similar to metal currency in the world, it can play the role of an international settlement currency, and can be held by governments and central banks as foreign exchange stabilization funds to intervene in the foreign exchange market.

Tavlas (1997) noted that when a currency acts as a unit of account, medium of exchange and store of value in international transactions without the participation of the currency issuing countries, that constitutes currency internationalization.

Mundell (2002) explained that when the circulation of money goes beyond the scope of the statutory circulation area, or fractions or multiples of the currency are imitated by other regions, that currency is internationalized.

The notion of currency internationalization discussed by the IMF is: the process of a currency crossing the boundaries of that country, becoming freely convertible, and getting traded and distributed in the world.

Sato (1999) presented the definition of internationalization of the yen by the Japanese Ministry of Finance: increase the proportion of non-residents' holdings of yen-denominated assets, especially the role in the international monetary system and to improve usage of yen in current transactions, capital transactions and foreign exchange reserves position.

Along with the increasing importance of RMB in international trading, internationalization of the RMB has also caused lots of concern in China. According to Zhong Wei (2002), internationalization of the RMB is a robust output of RMB-denominated financial assets and the process of the formation of an offshore RMB centre with the requisite liquidity. Tao Shigui (2009) believes that internationalization of the RMB is essentially the process to be a convertible currency as defined by the IMF.

A currency is affected by the entire economy where it domiciles and as an economic and social medium, currency facilitates all kinds of trading activities such as commodities and labor services.

When a currency starts circulating outside its own country, and acts as a universal equivalent in the world market, it functions as an international currency. When precious metals are circulating, international currencies function by original gold and silver bars based on actual weight. In the current international currency system, currencies of some economically developed countries can be converted freely and have stable values. They are accepted in trade and replace gold to perform the functions of world currency (international currency), such as the GBP, USD and EUR. At present, CNY has some stability and has been used as an external pricing and paying tool to some extent. At the end of 1996, it realized convertibility of current items, but there is still a big gap.

5.1.1.1   *International transaction medium and international payment means*

As a transaction medium, use of paper currency in domestic markets has greatly reduced the cost of economic operations. Similarly, use of international currencies has greatly reduced the transaction cost in international trade and investment. In addition, international currency also functions as a medium for international payments. For example, international debt, international aid and war indemnity need to be paid with international currency.

5.1.1.2   *International pricing unit*

International currency can be used as a pricing unit for commodities or assets in the international market. Currency is a value measuring unit. If the currency varieties are too many, that can cause confusion and increase the transaction cost. Therefore, commodity or assets transactions between two countries require an international currency recognized by both countries as a pricing unit for payment. In addition to this, international currency also prices currencies of other sovereign countries. As long as the exchange rate between currencies of other sovereign countries and a certain international currency is determined, the exchange rate between the currencies of the two sovereign countries in the world can be easily obtained.

5.1.1.3 Function of international value storage

The function of international value storage of international currency is generally reflected as foreign exchange reserves of various countries. For the sake of balancing international payments, stabilizing the exchange rate of local currency and providing guarantee for international debt, every sovereign country usually needs to have some foreign exchange reserves. Generally speaking, foreign exchange reserves are held in a key international currency. To preserve and increase the value of foreign exchange reserves, countries that have foreign exchange reserves usually use them to buy financial assets such as debt and stock issued in international currency issuing countries.

5.1.2 *Internationalization of currency function*

Currency internationalization is a process that is expected to make a country's currency acceptable across the borders, circulate overseas and become an internationally recognized pricing unit, settlement currency and storage currency. "Internationalization" usually means an on-going and unfinished process. As a process, currency internationalization has colorful connotations and extensive influences. First, currency internationalization is a process for extending the currency's domestic functions to overseas. Second, currency internationalization implies a process towards the currency's international acceptance. Third, currency internationalization is a process for a country's soft power to be strengthened. Finally, currency internationalization is a process for readjustment of the existing international currency pattern.

Currency internationalization has three levels: first, the periphery, which refers to circulation in neighboring countries and regions. Second, currency regionalization, which refers to a pricing and settling currency for regional transactions and investment, besides it being held in as an international reserve asset, or a new and uniform currency (such as EUR) for an international region integrated after long-term currency cooperation among various countries in an international region. Third, currency globalization, which refers to further extending currency circulation for extensive application and acceptance throughout the world.

Currency internationalization can be either partial or complete internationalization. Partial internationalization means that a country's currency performs only one of the following four functions (transaction medium, pricing unit, payment means and value storage) in international economy. Full internationalization means the currency performs all the four functions.

5.2 Internationalization of RMB

5.2.1 *Status of CNY internationalization*

The rapid growth of China's foreign trade and continual stability of CNY exchange rate in recent years, the international reputation of CNY is improving rapidly, and China's neighboring countries have begun to recognize and accept CNY as a transaction currency and an international settlement means. CNY internationalization has achieved initial success as follows:

5.2.1.1   *Sustained growth of foreign trade settlement in CNY*

In July 2009, Measures for the Administration of Pilot CNY Settlement in Cross-border Trade was announced by the People's Bank of China, the Ministry of Commerce, the Ministry of Finance, the China Banking Regulatory Commission, the General Administration of Customs and State Administration of Foreign Exchange. Since then, CNY can be used as a settlement currency for China's foreign trade. In 2010, CNY settlement in cross-border trade totaled CNY 506.3 billion, and the number of export enterprises involved in pilots increased from 365 to 67,000 at the end of 2010. In 2012, CNY settlement in cross-border trade in China totaled CNY 2,940 billion, and 193 countries and regions began CNY trade settlement with China. This helps China to further expand the acceptance of CNY beyond its borders and drive CNY to become an international trade settlement and trade financing currency.

5.2.1.2   *Continual expansion of currency swap agreement*

Since 2008, the People's Bank of China has signed currency swap agreements with many countries and regions. In December 2008, the People's Bank of China and the Bank of Korea signed a CNY swap agreement totaling CNY 180 billion. In January 2009, the People's Bank of China and Hong Kong signed a CNY swap agreement totaling CNY 200 billion. After that, it signed currency swap agreements with Central Bank of Malaysia and Central Bank of Argentina. By the end of 2012, China had signed currency swap agreements with 17 countries and regions (Table 5.1).

5.2.1.3   *CNY has begun to hold a dominant position in some of Chinese border trade*

At present, China has bilateral settlement and cooperation agreements with Russia, Mongolia, Vietnam, etc., confirming that commercial banks of both parties can open correspondent accounts and implement domestic currency settlement. At present, the settlement amount of CNY in Sino-Vietnamese trade is about 81 percent while the settlement amount of CNY in Sino-Mongolian trade is about 90 percent. CNY is becoming the main settlement currency in the trade between China and its neighboring countries.

5.2.1.4   *Multiple modes support external output of CNY*

CNY internationalization implies that CNY should have multiple channels to go abroad, flow to international markets and possibly become an international settlement currency, pricing currency and reserve currency. In addition to CNY

Table 5.1 Currency swap arrangements with China

	Countries/ Regions	Swap quota
2008	Korea	180 Billion RMB
2009	Hong Kong	200 Billion RMB
2009	Malaysia	80 Billion RMB
2009	Belarus	20 Billion RMB
2009	Indonesia	100 Billion RMB
2010	Iceland	3.5 Billion RMB
2010	Singapore	150 Billion RMB
2011	New Zealand	25 Billion RMB
2011	Uzbekistan	0.7 Billion RMB
2011	Mongolia	5 Billion RMB
2011	Kazakhstan	7 Billion RMB
2011	Thailand	70 Billion RMB
2011	Pakistan	10 Billion RMB
2012	UAE	35 Billion RMB
2012	Turkey	10 Billion RMB
2012	Australia	200 Billion RMB
2012	Ukraine	15 Billion RMB

cross-border trade settlement, there are several channels, such as consumption of overseas Chinese residents, tourist expenditure and activities of overseas and informal financial organizations.

5.2.1.5   CNY begins to settle in Hong Kong and Taiwan

In July 2010, the People's Bank of China and Hong Kong Monetary Authority signed a new settlement agreement which allowed enterprises and individuals to conduct CNY transfer and payment via banks. Since then, CNH has begun to develop rapidly. Till the end of 2012, CNY deposits in Hong Kong were up to CNY 630 billion, which indicated that the CNY capital pool in Hong Kong had begun to take shape. With further economic and trade exchange cross-Straits, CNY began to serve as a transaction medium and a payment means in Jinmen and Matsu. In January 2012, the People's Bank of China and Taipei Branch of Bank of China signed a CNY business clearing agreement under which Taipei Branch of Bank of China can provide participant banks with CNY clearing and settlement services according to the authorization. After that, CNY business in Taiwan began to develop quickly. According to Taiwan's official statistics, CNY deposits in Taiwan as of 31 March 2012 were up to CNY 50 billion.

5.2.2   Restraining factors of CNY internationalization

Although CNY internationalization has made significant progress, there are still restraining factors:

5.2.2.1   The development level of a financial market in China is low

If a country's currency wants to become an international currency, there must be a financial market which features a breadth (having multiple financial instruments) and a depth (a developed secondary market). A highly developed and open financial market can provide financial traders with a lot of transaction tools to meet the requirements of different investors on security, liquidity and profitability, which can help internationalization of the country's currency. However, the development of a financial market in China falls behind on this count and cannot be compared with developed countries in terms of volume of financial assets and transactions and financial innovation. These factors have severely hindered CNY internationalization.

5.2.2.2   The interest rate and exchange rate in China are not marketized as yet

If a country's currency wants to become an international currency, the interest rate and exchange rate of the country should be determined by market forces. There should be no distortion and suppressing. A market determined interest rate can reflect the real status of the capital market, and the fluctuation of the interest rate influences prices of financial assets, which makes the currency debtors and creditors follow market laws. Exchange rate is the conversion rate between domestic currency and foreign currency. The marketized exchange rate can truly reflect the relationship in a foreign exchange market and finally optimize resource allocation. At present, the Chinese interest rate and exchange rates have not been marketized as yet.

5.2.2.3   The financial regulation level in China falls behind

After a country's currency is internationalized, the financial risk for the country increases. This requires its government to have a very strong capacity for preventing financial risk. A strong risk prevention capacity means a sound financial supervision system, rich financial regulation skills and experience. However, the financial supervision system in China is still imperfect and the financial supervision experience is insufficient, which makes the financial supervision departments in China fail to effectively control overseas CNY.

5.2.2.4   Capital account is not open yet

With further advancement of CNY internationalization, CNY demand of Chinese non-residents will increase radically, which is conducive to CNY internationalization. To meet this demand and facilitate free entry and exit of international capital in Chinese financial markets, China needs free convertibility on capital account to make CNY capital an investment tool for international investors and promote the process of CNY internationalization. Although free convertibility of CNY on current account has been realized in China, free convertibility has not been realized for many capital items. Of the 40 capital items specified by the International Monetary Fund (IMF), only 14 are basically convertible in China. There are 22 capital items that are partially convertible. There are 4 capital items that are not convertible. This indicates that the convertibility on capital accounts in China is relatively poor. It can be said that the control of capital accounts is one of the greatest barriers for CNY internationalization at present.

It can be seen from the above analysis that CNY internationalization has made some progress, but there are still many restraining factors. In particular, the financial market in China lacks breadth and depth, and capital account is not freely convertible as yet. These two restraining factors cannot be solved in a short period. Then, some scholars have proposed in recent years that these two restraining factors can be overcome by establishing and developing an offshore CNY market.

5.3 Offshore financial market

5.3.1 Concept definition of offshore finance

Offshore financial markets serve non-residents who operate in convertible currencies and conduct financing in and out of the currency issuing country and is basically not limited by laws and tax systems. It is also called external financial market or Euro currency market. At present, this kind of new financial business has become an important part of the international financial system. (see Figure 5.1.)

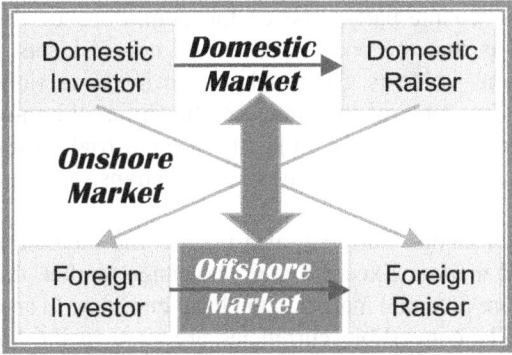

Figure 5.1 Domestic market and offshore market

Euro currency has become a classical pronoun for offshore currency. The currency standard in the definition of offshore finance is neglected. Overseas currency can hardly be the standard for distinguishing offshore from onshore. "Offshore" is not only limited to state borders but is a virtual "shore". Offshore finance is a systemic concept but not a regional concept.

Taking the USD in Europe for example, non-residents do not finalize transactions and investments with USD through financial institutions and financial markets in the United States but realize them through offshore financial centres beyond the US such as USD markets in London and Europe. Until the end of 2008, USD deposits of American non-residents in and out of the United States were USD 809 billion and USD 2,580 billion, respectively, occupying 24 percent and 76 percent of the USD market in the United States. This indicates that the USD of non-American residents is mainly saved in overseas market. At the end of 2008, non-residents issued bonds worth USD 3,640 billion of which USD 920 billion was held by American residents and USD 2,740 billion was held by non-residents. In the transaction, USD 3,200 billion was finalized through offshore markets.

5.3.2   *Characteristics of offshore financial markets*

Although the offshore financial market has developed over several decades, its connotation keeps changing. As a new market environment system that is totally different from an onshore financial market, the offshore financial market has its special characteristics:

5.3.2.1   *Internationalization is the primary characteristic of offshore financial markets*

Compared with onshore financial markets, the main characteristic of an offshore financial market is to make itself internationalized. This kind of internationalization can be reflected in many aspects. It is very thorough from counterparty, transaction currency, asset price, financing source and application etc.

First, from the counterparty, one of the definitions of the initial offshore financial market is that the counterparties are non-residents. With the market development, some residents also get involved in transactions in the offshore financial market, but most of the counterparties are still non-residents. A lot of international financial institutions, multinational industrial consortiums and capital accounts of various national governments are the transaction subjects in offshore financial markets.

Second, the transaction currency also reflects the internationalization of the offshore financial market. Except foreign exchange market, the transaction currency in an onshore financial market is the local currency. In an offshore financial market, universally convertible currencies are used in the transactions. Since both parties are non-residents, they can freely select the types of transaction currency according to their actual requirements.

Third, it means internationalization of assets price. Due to the openness and the non-residents operating in offshore financial markets, asset prices of offshore financial markets in the world are almost the same. The major reason is that these markets do not have more control than onshore markets. Therefore, offshore financial markets also shows internationalization in exchange rate, interest rate and assets price, etc.

Finally, it is internationalization of sources and applications of money. Except the political hedging capital, the capital source of an offshore financial market is usually hot money and working capital from all over the world. Profits are pursued worldwide. Therefore, the financing source and application in offshore financial markets reflect internationalization.

5.3.2.2   Few financial controls are the important characteristics of offshore financial market

In offshore financial markets, free finance is advocated; there are few financial controls. Offshore markets feature few controls in the currency issuing country, the country where the offshore market is located. It can be specifically reflected in interest rate, exchange rate, foreign exchange flows, deposit reserves and deposit insurance requirements. In addition, low-tax policy is also one of its characteristics. The reason is that the country where the offshore financial market is located reduces all kinds of tax rates so as to attract transactions, which expands the transaction scale and strengthens the international competiveness of the offshore financial market of the country.

5.3.2.3   Systematic form of market

Generally speaking, an offshore financial market is a kind of financial business transaction platform but not a physical market in a traditional sense. Offshore markets finalize transactions via electronic media. In fact, it is only a presentation of a special account.

5.3.2.4   Based on wholesale transactions, the asset liquidity is strong and the settlements are particular

Transactions in offshore financial markets are mainly by international large-scale financial institutions, dominant financing institutions of various countries and large-scale multinational companies. The transactions of these institutions are characterized by large sizes. The main transaction varieties in offshore financial markets are foreign exchange transactions. Moreover, transactions of interbank-offered credit occupy a dominant position and these are transactions among institutions. Therefore, offshore financial markets generally have "wholesale" transactions, not retail transactions between institutions and individuals. Due to the characteristics of interbank offers, its asset liquidity is relatively high. Since the currencies transacted in offshore financial markets are multiple convertible currencies, their modes of settlement are different from the domestic financial market. Free payoff services need to be provided to offshore counterparties all over the world.

In conclusion, these characteristics of offshore financial markets are interconnected. Since the purpose of offshore market development is to attract international working capital to be transacted in the market, the offshore financial market must be internationalized. To realize this internationalization, few financial controls and restrictions and low tax rates are used to attract the inflow of international capital. It is matched with advanced electronic and paperless transaction systems. Similarly, it also requires infrastructure construction which facilitates offshore financial transactions. As a lot of international working capital flows into an offshore financial market for transactions, its inherent large-scale and convenient transaction characteristics make the offshore financial market obtain such characteristics as wholesale dominance, strong asset liquidity and special settlement mode.

5.3.3   Development mode of offshore financial markets

The current development mode of global offshore financial markets shows that it has unique characteristics in business and operation scope, market formation causes and market functions. In particular, the most typical and common characteristic that distinguishes it from the domestic market is business scope. According to this standard, the existing offshore financial market can be divided into separated type, mixed type, tax haven type and penetrating and separated type.

The biggest characteristic of a separated-type offshore financial market is that separate accounts must be opened for offshore and traditional business. Moreover, the flow of working capital between the two accountants is strictly limited. After being approved by the financial authority, offshore business can be offered with no payment for deposit reserves and deposit premium as well as local tax immunity. The representative markets include the New York and Tokyo markets.

Besides offshore business, a mixed-type offshore financial market allows non-residents to operate onshore business. No financial authority needs to approve the offshore business but all bank reserves and relevant taxes must be paid. The currency transacted in the market is not the local or regional currency. Cross-border capital flows are not restricted. The representative markets include the Hong Kong and London markets.

The most typical characteristic of a tax-haven-type offshore financial market is that the local political situation is relatively stable, taxes are low or even non-existent. There are no actual offshore capital transactions in this market. Only account keeping for other financial market transactions is handled. When there are no financial controls, the offshore financial transactions can achieve the purpose of avoiding capital supervision and tax reduction and exemption. Therefore, it is reputed for being "a haven for tax avoidance". The representative markets include the markets in the Caribbean: the British Virgin Islands, Bahamas, Cayman Islands and Bermuda.

A penetrating- and separated-type offshore financial market is based on the separated-type market. The accounts for offshore and onshore businesses are opened separately. The difference lies in allowing capital interpenetrating within a certain scope. For example, residents are allowed to invest in the domestic market. The banking and financial institutions which operate offshore businesses can also lend capital in offshore accounts to domestic enterprises. The representative markets include the market in Singapore.

5.3.4 Development of an offshore market to support CNY internationalization

The supporting role of offshore markets for currency internationalization can cause the necessary conditions for currency internationalization to mature. Moreover, offshore markets can enhance the depth and breadth of currency internationalization (see Figure 5.2).

5.3.4.1 Demand for CNY transaction convenience

The necessary condition for a currency to become the main settlement currency in the world lies in use safety, convenience and strong liquidity. Moreover, it can be used for exchange, settlement, payment and investment and financing in an international currency market. The currency can be used for 24-hour seamless transactions worldwide. China covers five time zones, so it cannot meet the requirement of the convenience of global currency transactions simply by relying on Chinese local financial markets. However, the establishment of an offshore

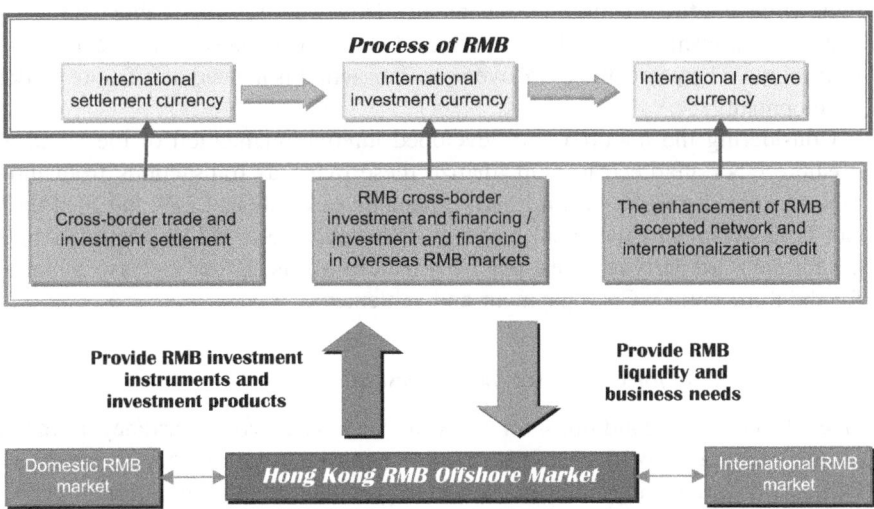

Figure 5.2 RMB offshore market in the process of RMB internationalization

CNY market beyond the borders can just solve the problem of 24-hour transactions in CNY.

From the time zone perspective, Hong Kong is the most important offshore CNY market and shares the same time as Beijing in China. It can be considered as an offshore extension of Chinese onshore financial markets. Singapore is also located in the Eastern Eighth Zone and can radiate throughout Southeast Asia and Oceania for retail of CNY offshore business after undertaking wholesale CNY offshore business from Hong Kong. It can play a helpful supplementary role with a Hong Kong offshore market. London has a seven-hour time difference with China. The closing of the Hong Kong offshore CNY market can just be linked up with the opening of the London offshore market. New York has a five-hour time difference with London. That is, the New York offshore market can still be open for transactions after the London market is closed.

From the geographical location perspective, the establishment of offshore CNY markets in Singapore, London, New York and Tokyo can radiate the main global economic entities from Asia, Europe, America and Oceania. Then, such problems as time zone connection, geographical distribution scope and matching of wholesale market and retail market can be solved effectively. Therefore, the development of an offshore CNY market can greatly enhance transaction convenience of CNY.

5.3.4.2   Demand for CNY overseas transactions

One of the goals of CNY internationalization is to make CNY a main world settlement and payment currency. Then, use of CNY for China-related transactions and direct investment will not be the main content of CNY transaction in the world. CNY internationalization will certainly make the overseas CNY transactions irrelevant to the Chinese real economy. However, it is impossible to perceive that these transactions will be conducted only by Chinese onshore financial institutions. Therefore, the CNY overseas transaction is in need of offshore market development.

Considering the matured and developed internationalization of the USD, an offshore CNY market can help finalize these overseas transactions from three aspects: first, it results in third party financing in offshore CNY and then there can be a swap into the local currency transaction. Second, it acts as a market for the third party to invest in CNY financial assets. Third, it acts as a market for the third party's CNY payment and settlement.

5.3.4.3   Demand for CNY assets held overseas

One of necessary conditions for CNY to become a world currency is that it should become a world reserve currency. Therefore, a lot of CNY assets should be held by foreign investors in securities products or reserve currency. As for these investors, security, liquidity and profitability of CNY assets are the important factors for them to select the varieties of reserve assets. In terms of security, the country where the offshore market is located should have a stable political environment and sound legal and tax systems and protect the investors' privacy well. In terms of liquidity, an offshore market does not have as many restricting measures as an onshore market. Moreover, the financial institutions operating in offshore markets are usually large-scale international banks, so they can provide investors of CNY assets satisfactory liquidity. In terms of profitability, since the competition in offshore markets is relatively sufficient, the financing cost is very low. It is suitable for investors to use the leverage to conduct trend investment. According to the experience of USD development, the foreign investors for USD will invest 70 percent of USD assets in offshore USD market.

China has had a significant trade surplus for many years. These surpluses are calculated in USD. This phenomenon is not conducive to the growth of overseas CNY assets. From this perspective, the construction of an offshore CNY market is very important. On the one hand, the offshore market can improve the money multiplier of overseas CNY, which can make undiversified CNY assets generate more assets. On the other hand, an offshore market also provides deeper and wider markets for CNY assets to go abroad.

5.3.4.4 Demand of CNY overseas circulation

One of the restraints for CNY to become an international currency is the impact of overseas assets on domestic currency after CNY internationalization. CNY internationalization can be realized only when there are sufficient CNY transactions and financial transactions. However, if huge transactions need to be finalized with domestic financial markets in China as a counterparty, it will make the currency policy-making of the Central Bank of China difficult and severely influence its impact. Then, the contradiction between CNY internationalization and currency policy seems unavoidable. However, if the flow quantity and scale of CNY overseas assets in and out of borders are reduced, it will greatly relieve this contradiction and will solve the contradiction between CNY internationalization and currency policy to some extent. This would imply that overseas CNY assets can have self-circulation beyond the borders. Specifically, the main currency transactions and financial transactions of overseas CNY assets need to be finalized beyond the borders. Naturally, offshore CNY will be used to finalize the transactions. Thus, the overseas cycle mechanism of CNY among offshore markets can keep CNY assets within a certain scale and any CNY transaction will not necessarily be finalized through domestic financial institutions. This will greatly reduce the impact of CNY internationalization on domestic currency policy and can effectively solve this contradiction.

5.4 The offshore RMB market in Hong Kong (CNH)

Offshore CNY finance refers to CNY financial transactions or financing conducted by non-residents beyond the financial supervision system and financial operation system of local authority.

Garber (2011) believes that expected appreciation of RMB is an important factor driving the development of the RMB offshore market in Hong Kong. The arbitrage trading logic is that expected appreciation of RMB has enhanced speculative demand in the Hong Kong RMB market, resulting in the difference between offshore and onshore RMB spot market widening, and leading to more Mainland importers choosing to settle purchases at the foreign exchange in Hong Kong. The end result is the increased supply of RMB deposits in Hong Kong and narrower spreads between offshore and onshore markets. He then said, because of the incentive of expected RMB appreciation, even real transaction-based cross-border RMB trade settlement and Foreign Direct Investment (FDI) may also have a speculative element.

Murase (2010) pointed out that the difference between the market for onshore and offshore RMB exchange rates is essentially a Tobin tax. Under the RMB expected appreciation, this spread is the cost paid by foreign funds when they transfer capital into Mainland China. But at the same time it is a subsidy for the output of domestic RMB.

Yu Yongding (2012) examined arbitrage between offshore and onshore markets based on the interest rate parity and pointed out that, in Hong Kong's offshore market, the main reason of RMB depreciation was the result of directional reversal of arbitrage activities in the second half of 2012.

Maziad and Kang (2012) applied GARCH for analysis and showed that the onshore market RMB spot prices had a significant impact on the offshore market and at the same time, forward RMB prices in the offshore market can be used to predict the forward exchange rate of the RMB onshore market. In addition, they also found the volatility spillovers between the two markets.

Wang (2011) pointed out that RMB deposits in Hong Kong have three characteristics and can be considered as hot money. First, the major holders are enterprises other than residents; second, the main motivations of holdings are expectation of RMB appreciation and interest rate spreads; and third, they are waiting for the opportunity to flow back.

At present, the "onshore" offshore financial form for CNY has not appeared as yet. China's border can still be treated as evidence for internal and external separation of financial supervision system and financial operation system of local authority.

Offshore CNY financial business mainly includes CNY deposits, exchange, trade financing, bond, stock, capital transaction, insurance, investment and assets management. When the scale of offshore CNY financial business, number of participants and active degree of transactions are up to some extent, an offshore CNY market will form.

CNH refers to an offshore CNY financing and transaction venue formed from offshore CNY business in Hong Kong. Since the offshore financial business of CNY is mainly in Hong Kong, CNH has become a transaction and a pricing centre for offshore CNY.

CNH is a policy-driven offshore financial market formed under the guidance of government policy in China. CNY and HKD belong to the same sovereign state. CNY belongs to an offshore financial market for domestic currency. Therefore, policy-driven formation and offshore finance of domestic currency are the main characteristics of CNH.

Meanwhile, CNH is a part of the Hong Kong offshore financial market. Early in the 1970s, the offshore financial market was formed in Hong Kong, taking USD as the dominant currency. CNH is a further expansion of the Hong Kong offshore financial market. Therefore, the Hong Kong offshore financial market includes offshore financial activities in traditional currencies such as USD and offshore JPY. As a new offshore currency, CNY's offshore financial activities will be an important part of the Hong Kong offshore financial market.

5.4.1 Establishment background of CNH

Aiming at the foreign and domestic market situations, CNH is not only feasible but also necessary under the comprehensive influence and function of multiple factors.

5.4.1.1 International environment

In the process of financial internationalization and liberalization countries and regions from Asia–Pacific have performed outstandingly. Although the construction and development of offshore financial centres show different characteristics in different countries and regions, the encouragement and policies from local governments is usually similar. Their purposes are to create good economic and political environments for offshore financial market construction and enhance their international position in local country or region.

There are many financial markets in China's neighboring countries and regions. Their active performance will influence the international capital flow. Construction of new financial centres is restricted by relevant domestic systems. Due to the relatively closed financial environment in China's domestic market, to develop offshore financial centres has become an important step for the Chinese economy to be internationalized. The development of traditional offshore financial markets in the Asia–Pacific region, such as the Singapore offshore financial centre, has matured gradually. The newly-built offshore financial markets such as the Malaysian market are beginning to take shape. From the previous example of successfully building offshore financial markets worldwide, it is possible to build offshore financial centres as long as we grasp the historical opportunities and make full use of our own advantages.

5.4.1.2 Domestic environment

In the twenty-first century, GDP, living standard of people and development of international economy and trade in China have made remarkable achievements. Neighboring Hong Kong, the Shenzhen Special Economic Zone has many foreign-funded enterprises and domestic enterprises which have developed overseas businesses, which puts forward urgent requirements for China conducting internal and external CNY business and foreign exchange business. China is a developing country, and its economic system cannot be liberalized fully. Interest rate and credit control, singularity of currency policy tools and foreign exchange controls have to continue until construction of efficient supervision systems is completed. These restraints result in a significant gap with the world leading offshore financial markets which feature little financial supervision, many preferential policies and high financial liberalization. Simply developing domestic offshore financial businesses cannot put China's economy on par with advanced countries. On the other hand, the Hong Kong Special Administrative Region (HKSAR) is an old and established international offshore financial centre in Asia and has a more mature financial system than China's domestic market, so it will logically become the first region for China to develop CNY offshore business. CNH establishment does not only open up a convenient door for domestic and foreign-funded enterprises to invest in domestic and international markets but also boosts China's key and tentative steps towards internationalization of its currency while continuing with controls on the domestic market and financial system.

5.4.1.3   *CNY internationalization determines the market positioning*

Compared with mature European currency market, CNH is not a genuine and complete offshore market. Hong Kong offshore CNY business needs to be enriched and expanded. CNH development can promote CNY internationalization which determines the functional positioning of a Hong Kong offshore market. China has had trade surplus for a long time, and it must enlarge overseas circulation of CNY in order to avoid undertaking the adjustment cost caused by international financial market unrest and get rid of the positioning constraints vis-à-vis international currencies such as the USD. As one of the global offshore financial centres, the Hong Kong market should keep developing CNY settlement, clearing and payment businesses. On the premise that the domestic capital items have not been completely opened and that the risk of opening them is still huge at the present stage and upon the current global economic adjustment, the Hong Kong offshore market should be developed into a buffering platform for offshore CNY so as to weaken the direct impact of international market uncertainties on the domestic market. The continual advancement of CNY internationalization requires not only continuous growth of CNH but also ensuring its development direction. Both of them promote each other and coexist.

5.4.2   *Advantages for Hong Kong to establish CNY offshore centre*

5.4.2.1   *Superior natural condition*

On the one hand, Hong Kong is located in the heartland of the Asia–Pacific region. Since ancient times, it has been treated as a free trade port and keeps close economic relationship with Mainland China and European and American countries. On the other hand, the time zone of Hong Kong is between Euramerica, which effectively connects the financial transaction activities of European and American economic entities and plays an important role in smoothing global

24-hour transactions via various financial instruments. Besides that, Hong Kong and Mainland China are located in the same time zone and can communicate with each other. It has good recognition for the Mainland and remains a close economic, political and cultural relationship with it. These natural advantages for Hong Kong market lay a solid foundation for it to become a CNY offshore market.

5.4.2.2 Hong Kong has rich experience and foundation for offshore financial development

In the early 1970s, the offshore financial market was born in Hong Kong. In 1978, Hong Kong's overseas loans occupied 54.4 percent of total loans, and its overseas liabilities occupied 42 percent of total liabilities. In 1980, 88 of the 115 licensed banks were foreign-funded. This developed international financial business has made Hong Kong a very important offshore financial market. The assets of the Hong Kong offshore financial market grew from USD 17.5 billion in 1977 to USD 101.2 billion in 1985. After development over 30 years, Hong Kong has become one of the globally known offshore financial centres. At the end of 2010, assets of all authorized institutions in Hong Kong totaled HKD 12,296 billion of which foreign currency assets were HKD 7,552 billion or 61.42 percent.

The management and service levels of Hong Kong offshore finance are relatively high. As of December 2010, there were 146 licensed banks in Hong Kong of which 23 were registered in Hong Kong and 123 were registered out of Hong Kong. There were 199 foreign-funded banks in Hong Kong, including 70 out of the biggest 100 banks in the world. Many foreign-funded banks form an important basis for the development of the Hong Kong offshore financial market. The Hong Kong offshore financial market is of the internal and external mixing type, and there is no strict division for local residents and non-residents. Legally, there are no special preferences or supervision for offshore financial business. Local and foreign-funded banks conduct fair competition and participate in all kinds of financial activities on an equal basis.

5.4.2.3 Developed and sound financial system

As a leading international financial centre, the Hong Kong government's continual reform and innovation have accomplished a highly developed financial system and an attractive business environment.

Hong Kong is the twelfth largest banking centre in the world, having a perfect financial system, developed financial infrastructure and strong anti-risk capacity. After experiencing the Asian financial storm and the global financial crisis in 2008, its capital market has become even more matured. The stock market and bond market keep expanding, and the fundraising function of capital market is becoming increasingly stronger. At present, it dominates the Hong Kong financial system. The gold market of Hong Kong is one of the four famous gold markets in the world, and its geographical location can link up with time zones of London and New York. It is indispensable for the global gold market. Hong Kong has a local London gold market. These unique advantages have attracted many international gold tycoons to set up gold investment businesses in Hong Kong, which drives Hong Kong to be one of the most developed gold markets. Since it has the pegged exchange rate system and there are no foreign exchange controls, the openness of the Hong Kong foreign exchange market is very high and features low foreign exchange risk and free circulation. Its huge transaction volumes give Hong Kong important influence in global foreign exchange markets. Since the international financial crisis in 2009, the Hong Kong financial system has successfully resisted the impact, and its performance in the crisis reflects its stability and development.

5.4.2.4   *High financial liberalization in Hong Kong*

Stock, foreign exchange, futures and gold can enter Hong Kong and be sold. In January 1973, Hong Kong terminated foreign exchange and gold transaction control. In March 1978, it terminated the restriction that foreign-funded banks were not allowed to enter Hong Kong to establish branches. In March 1982, it canceled the regulation that foreign currency deposits were collected with 15 percent interest withholding tax. Due to the preference and slack policies in tax and financial control in Hong Kong, the financial liberalization in Hong Kong is very high. In a relatively free financial environment, many foreign-funded banks have set up shops in Hong Kong and have formed an important foundation for the development of the Hong Kong financial market. At the end of 1961, there were only 23 foreign-funded banks out of 85 licensed banks in Hong Kong. In 1975, the number of foreign-funded banks was only 28. After terminating the restrictions on foreign-funded banks in 1978, the foreign-funded and licensed banks increased rapidly. In 1988, there were 88 foreign-funded banks. In 1990, there were 138. In 2010, 123 (84.25 percent) of the 146 licensed banks were foreign-funded. High financial liberalization in Hong Kong also provides an important condition for the development of Chinese-funded banks in Hong Kong. In 2010, there were 14 Chinese-funded banks, 11 UK-funded banks, 10 Japanese-funded banks and 9 US-funded banks out of 146 licensed banks. There were 10 local Hong Kong banks. The competiveness and market share of local Hong Kong banks were in a sluggish state.

5.4.2.5   *Hong Kong has a perfect financial system and supervision system*

The Hong Kong government abides by the principle of free market and does not interfere with the financial market operation to the extent possible. There is no foreign exchange and capital controls in Hong Kong, so foreign companies can participate in the local financial market freely. The main supervision authorities in Hong Kong include Hong Kong Monetary Authority (HKMA), Securities and Futures Commission (SFC), Office of the Commissioner of Insurance (OCI) and

Mandatory Provident Fund Schemes Authority (MPFA) which are responsible for supervising the banking industry, securities and futures industry, inductance industry and retirement plans, respectively.

The main purpose of HKMA is to function as the central bank by maintaining financial systems and stability of banking industry. Besides banking industry supervision, HKMA is also responsible for maintaining HKD stability, improving financial system efficiency, promoting its development and protecting honest and just financial system. SFC is responsible for implementing and regulating the legislations for the Hong Kong securities and futures market, promoting and boosting the development of the securities and futures market, regulating and supervising Hong Kong Exchanges and Clearing Limited and its subsidiaries including the Stock Exchange of Hong Kong Ltd., Hong Kong Futures Exchange Limited and three recognized clearing companies. OCI is responsible for implementing the legislations for insurance companies and intermediaries.

Meanwhile, a three-level bank system is implemented in Hong Kong, which can strengthen bank supervision and management and promote a sound bank supervisory system. Therefore, although Hong Kong has preferential tax and systems, its perfect legal and supervisory systems have realized a stable and healthy financial environment.

5.4.2.6 Good development environment formed from mutual dependence with domestic economy and strong support of central government

The differences between advantages and disadvantages of Mainland China and Hong Kong have formed a basis for mutual advantage and complementarity and interdependence. The market of Mainland China is huge, there are rich product factor resources such as manpower, raw materials and energy, the cost is low, and there is a profound economic base. However, it is restricted by huge population, insufficient development capital and shortage of advanced technology, international marketing and experience. On the contrary, Hong Kong has the financial industry as its powerful advantage. It has strong financing capacity and takes the lead in scientific research in the world. As an international financial centre, Hong Kong has accumulated rich experience in international marketing. However, it is restricted by the regional area, the cost of production factors such as land is expensive, internal market is small, economy of scale is not available, the development space for local enterprises is limited, industrial upgrading is slow and its resistance to risks is weak. The economic integration between Mainland China and Hong Kong has promoted the rearrangement of resources, eliminated the barriers in the flow of production factors, accelerated the flow of commodity, capital, technology and manpower, further highlighting the comparative advantages of the products and has greatly strengthened the international competiveness of products. At the same time, it has also driven Hong Kong and Mainland China to positively adjust the industrial structure and strengthen the economic power.

5.4.3   Development status of offshore CNY market

Statistics of the Society for World Inter-bank Financial Telecommunications (SWIFT) show that in August 2015 the market share of CNY had reached 2.79 percent, surpassing the yen and becoming the world's fourth largest payment currency. The higher status of CNY is closely related to the role of the Hong Kong CNY offshore financial center. According to statistics, about 60 percent of foreign direct investment in the Chinese Mainland comes from Hong Kong, and about 60 percent of the foreign direct investment flows to Hong Kong; about 50 percent of foreign indirect investment (securities investment) flows to Hong Kong; in the Chinese Mainland about a quarter of the international trade passes through Hong Kong in the form of offshore trade or transit trade. Today, after more than 30 years of reforms and opening up, Hong Kong is still the important portal of international trade and investment in the Chinese Mainland. The construction of a Hong Kong offshore renminbi center is an important component of the CNY internationalization strategy, so it has the attention and support of the central government.

CNY deposit business has developed strongly in Hong Kong, and has a very close relationship with CNY settlement business for cross-border trade and investment in China. In the Hong Kong customer deposit market, the proportion of offshore CNY deposits increased from 1.04 percent in 2008 to 11.16 percent at the end of August 2015. CNY has become Hong Kong's third largest deposit money.

The Hong Kong CNY bond market started late but developed very fast. Issuance of CNY denominated bonds in Hong Kong is significant. Offshore CNY bonds of 1–3 years, 3–5 years and less than 1 year constitute a large proportion of the bond market in Hong Kong. At the end of September 2015, their proportion was 70.59 percent, 44.03 percent and 48.30 percent, respectively. At the same time, issuers tend to be diversified, including the Ministry of Finance, policy banks, financial institutions and non-financial enterprises. The depth, breadth and openness of Hong Kong's foreign exchange market are higher than onshore foreign exchange markets. The exchange rate pricing has a good market foundation.

In the future, the offshore CNY market in Hong Kong has to overcome some obstacles. The most critical problem is the mechanism of contact between offshore and onshore markets (channel of interest arbitrage), because without interest arbitrage, the market will continue to lack adequate competition and lose its vitality and efficiency.

Lack of an interest arbitrage mechanism between the markets can have serious influence on pricing efficiency; the phenomenon of assets with the same quality but different prices is relatively prominent. People have great hopes for Shanghai-Hong Kong Stock Connect (and Shenzhen-Hong Kong Stock Connect later) since it is being viewed as an important way to use offshore CNY funds. However, the actual effect is not as good as expected, and the transactions on both sides are asymmetric.

5.4.4 Positive roles of development of the CNY offshore centre in Hong Kong

5.4.4.1 Provide an experimental platform for CNY internationalization

The construction of an offshore financial centre is an important strategic step for facilitating CNY internationalization. As the pressure of CNY appreciation is sustaining, a window has to be found for experiments. As the financially most open and liberalized region, Hong Kong has advanced risk monitoring systems in place and should be the best choice for CNY offshore pilot. At present, offshore CNY can indirectly integrate liberalization on capital account and the internationalization process as a whole in Hong Kong, with the CNY–HKD–other hard currencies route and HKD bearing the transitional link. This implies that capital account transfers are basically open. Hong Kong thus becomes an overseas CNY deposit and loan settlement centre, which is similar to the way London became an overseas USD centre. This indicates that CNY is actually internationalized, centering in Hong Kong. To some extent, the process of establishing a CNY offshore centre in Hong Kong can be used as a reference for a CNY reform timetable, coordinating with CNY reform simultaneously and expanding businesses gradually. This will test and prepare for final opening of CNY.

5.4.4.2 Facilitate the reform of CNY interest rate marketization

The development of an offshore CNY market in Hong Kong can lead to a completely packetized CNY interest rate index. Currently, interest rate marketization is quite limited but the development of a Chinese financial market requires an interest rate index based on marketization as a reference for each financial decision. If the open markets of Hong Kong are fully used, a marketized interest rate index can be formulated by absorbing CNY circulating overseas, which will have a positive value for Chinese CNY interest rate marketization and the development of financial markets. The CNY offshore deposit interest rate or other interest rates in the free Hong Kong market can better reflect the commercial risks and can provide an effective reference for the formation of a CNY interest rate in the Chinese Mainland. At present, London Interbank Offered Rate (LIBOR) formed in the London offshore market is still an important reference for the formation of the USD interest rate, which just indicates that the offshore centre can play an important role in the formation of interest rates of the domestic currency.

5.4.4.3 Provide reference for domestic foreign exchange market

Hong Kong as a CNY offshore centre can provide a reference for the foreign exchange market in the Chinese Mainland. At present, the foreign exchange market in the Chinese Mainland lacks sufficient financial instruments that can avoid foreign exchange risks. The adjustment of foreign exchange policy by the central bank certainly lacks sufficient market basis. For the Chinese Mainland, the positive significance of a CNY offshore market lies in the following: if CNY is allowed to circulate in Hong Kong, one important benefit for supervisory authority is to include CNY circulating overseas in the banking system so as to facilitate it to master the changes of CNY's overseas flows and then take the corresponding measures.

5.4.4.4 Promote economic cooperation between China and Asia–Pacific

The development of an offshore CNY market in Hong Kong can drive the depth of economic cooperation between China and the Asia–Pacific region. The Chinese government is advocating a regional free trade zone framework in Asia under which economic relations among various countries in the region will be closer. If CNY is internationalized enough, competitors who price goods and services in USD and contend with export markets of developed countries can start pricing in CNY their exports to the Chinese market.

5.4.5 Main barriers of developing an offshore financial market in Hong Kong

Although offshore CNY market business in Hong Kong has developed significantly within a few years, Hong Kong still has a long way to go before it develops into a mature offshore CNY market. The reason for this is that many factors are restraining the development of CNH, such as systemic restraints, money laundering and an imperfect overseas CNY backflow mechanism. If no measures are taken for these factors, the development of CNH will remain greatly restricted.

5.4.5.1 CNY backflow mechanism in Hong Kong is not consummate yet

The backflow channels of CNY in Hong Kong have failed to meet the real requirements. To increase CNY backflows a perfect backflow mechanism has become an issue that must be solved for development of the Hong Kong offshore market. Currently, CNY circulating in Hong Kong cannot effectively flow back to the Chinese Mainland. CNY in Hong Kong is mainly absorbed by three types of overseas institutions, namely, overseas central banks, clearing banks of Hong Kong and Macau and overseas participant banks. Most of CNY (over 80 percent) in offshore markets are saved in Hong Kong banks which save it in BOC (Bank of China) Hong Kong that clears CNY capital in offshore market. At last, BOC Hong Kong saves CNY in the Shenzhen Branch of the People's Bank of China. Therefore, the mode and channel for CNY capital backflow to the Chinese Mainland, from the offshore market, are limited to cash. At present, the measures for Qualified Foreign Institutional Investors (QFII) have just been implemented, aiming at increasing backflow channels for overseas CNY. The construction of a CNY offshore centre is mainly dependent on a deep CNY capital pool and various CNY-based investment products. A perfect and efficient backflow mechanism can not only lead the increasing CNY capital pool in Hong Kong to the huge domestic market but also promote the market interaction between Hong Kong and the Chinese Mainland. Certainly, since the offshore financial market is greatly influenced by the policies in Mainland China, to make CNY backflow mechanism more sound and perfect is greatly dependent upon the relevant policies in the Chinese Mainland.

5.4.5.2 Bottleneck for inventory expansion of CNY in Hong Kong

To construct a highly mature offshore market, CNY assets in the Hong Kong market must be further increased so as to stabilize and support market operation. In 2011, CNY deposits in Hong Kong were increasing but in January–April 2012 CNY deposits in Hong Kong kept declining. At the end of December 2011, offshore CNY deposits were CNY 588.529 billion, which was far away from the expected CNY 1,000 billion. This triggered doubts about the prospects of CNY offshore market in the industry.

When the appreciation anticipation (important driving force) is lost, the inventory of CNY in Hong Kong keeps declining which directly restricts CNH. At this moment, measures can be taken to develop new CNY investment channels. The confidence for CNY holding volume in Hong Kong industry can be strengthened through high earnings, which can change the declining trend of CNY inventory.

5.4.5.3 It is restricted by system and money laundering will appear easily

The greatest systemic bottleneck of CNH development lies in Hong Kong. Although Hong Kong implements a "one-country-two-system policy", CNH is not different from CNY in a pure sense. First, CNY business in Hong Kong is not beyond the control of the People's Bank of China, and it is developing under the strict monitoring of the central government, the People's Bank of China and HKMA. Second, clearing of CNY business in Hong Kong is conducted by BOC Hong Kong, under Bank of China. Third, under the framework of the "one-country-two-system policy", Hong Kong has a high degree of autonomy, but the influence of the Chinese Mainland on Hong Kong is substantial. Therefore, it is not possible that Hong Kong is a fully independent and pure offshore financial market, for the Hong Kong offshore financial market is greatly influenced by the policies of the Chinese Mainland.

Furthermore, offshore financial markets respect client privacy, which provides the fertile soil required for money laundering. As CNH grows, transaction volume of CNY capital will increase, its scale will be larger, and more capital will be flowing frequently between the markets in Hong Kong and Mainland China. This will doubtlessly provide the perfect system for money laundering, which influences long-term and stable economic development in Mainland China and Hong Kong.

5.4.6   Suggestions on the development policy of Hong Kong CNH market

Great restrictions limit the supply and backflow of Hong Kong RMB, especially supply of overseas RMB. Therefore, a larger supply of overseas RMB, especially Hong Kong RMB, and fewer obstacles to the backflow are critical to further development of the Hong Kong CNH market and elimination of restrictions on RMB cross-border flow in the next several years.

5.4.6.1   Deregulate the RMB capital flow between the Mainland and Hong Kong under the guarantee of putting risk monitoring and control measures in place

In terms of RMB export, the following policies can be considered: (1) fully releasing the RMB settlement business in cross-border trade; (2) permitting Hong Kong banks to join the national inter-banks' lending market, but conditionally permitting RMB lending to Mainland banks; (3) allowing foreign enterprises to obtain RMB through currency swap agreements between China and other countries or regions and to invest in the Hong Kong market; (4) allowing the approved domestic enterprises' foreign investment in RMB; (5) allowing QDIIs to participate in foreign issues and trade in RMB.

In terms of RMB backflow, there are some available polices: (1) conditionally permitting Hong Kong banks to issue RMB loans to Mainland residents; (2) cancelling the export RMB settlement restrictions on regions and enterprises. For that, coordination and adjustment of export policies should be accelerated to make the export RMB settlement policies unobstructed.

5.4.6.2   Payment in RMB as much as possible for international balance of payments of some capital accounts

The main channels include: (1) the capital accounts in the Mainland that have been open to overseas can choose RMB as the means of payment instead of USD or other foreign currencies; (2) the overseas loans and economic aid projects are issued in RMB as much as possible.

5.4.6.3   Construction of various RMB markets and improvement of RMB clearing system in cooperation with Hong Kong government to guarantee a wider channel for use of RMB

The Mainland and Hong Kong government should join hands to promote the construction of various RMB markets, and to fully open and encourage the existing Hong Kong financial markets to issue financial products valued and transacted in RMB. Measures include: (1) enhancing the creation of RMB financial products and business, and increasing the Hong Kong RMB loan-to-deposit ratio; (2) encouraging insurance companies and fund companies to launch insurance products and fund products in RMB; (3) further developing the CNH bond market, expanding the scale of debt, and completing the structure of the bond period; (4) launching investment hedges suitable for the process of opening the domestic capital accounts, such as RMB deliverable forward, interest rate swaps, currency futures, etc.

5.4.6.4   *Promotion of further integration of financial markets in Shenzhen and Hong Kong, and acceleration of symmetric extension of RMB cross-border flow*

Further market integration of the two regions can facilitate effective RMB flows, inward and outward, and improve the use efficiency of RMB capital. The following specific measures may be considered: (1) to strengthen the relationships among products traded in the two regions, including increasing common listing, attempting to develop exchange traded funds (ETFs), depository receipts, and other cross-border market derivatives in both regions; (2) to expand the mutual access between market participants and financial intermediaries, and widen the flow channel of capital.

5.5   Opportunities of Hong Kong in the Belt and Road strategy

Participation in the construction of the Belt and Road is an important opportunity for the development of Hong Kong. Hong Kong is located in the geographical centre of China–ASEAN. Due to the special geographical location, coupled with Hong Kong's close economic ties and connections with the main countries or regions of ASEAN, Hong Kong can function as an important node in the construction of the "Marine Silk Road". Hong Kong is the world's freest and most open economic system and its social and business environment is well integrated with the international culture.

In economic and trade exchanges with Western countries, Hong Kong has systemic advantages such as a capitalist economic system, international business network and international talent. It functions as a buffer that helps China contain proliferation of Western ideology, reduce the international risk of outgoing enterprises and provide international talent. Hong Kong adopts the international system and business rules, which can reduce the operational costs of the Belt and Road. Chinese enterprises face great risk when they go out, because they are not familiar with rules of the international business, foreign investment and trade. To enter Eurasia, the Belt and Road needs effective integration and cooperation of enterprises in Hong Kong and the Chinese Mainland. Aggregate competitiveness of enterprises in the Chinese Mainland investing in foreign countries shall be improved, and enterprises in the Chinese Mainland shall be assisted to effectively avoid foreign investment risks.

Hong Kong will also be able to provide international talent for the Belt and Road. Hong Kong has a large number of high-level professional international talents. It is leading in the world in finance, law, consulting, accounting and other aspects. Financial services in Hong Kong are high-end. Hong Kong lawyers are familiar with common litigation methods and skills. Hong Kong consultants are good at big data processing and precision analysis. Professionals generally have high foreign language competence and comprehensive understanding of international concepts.

To build the interconnected financial supporting system, Hong Kong can become an important financing platform also. The countries and regions along the Belt and Road are generally in the rising stage of economic development. The industrialization and urbanization process is accelerating, and infrastructure investment is in great demand. Hong Kong can provide investment and financing support for infrastructure construction, resource development, industry cooperation and other related projects of countries and regions along the Belt and Road. Hong Kong can help bridge the huge funding gap of different countries in the field of infrastructure construction, and provide customers with a variety of financing options, including offshore CNY financing. Hong Kong is the world's largest offshore CNY center at present, with the largest CNY working capital pool outside the Chinese Mainland.

6   Hong Kong gold market and its products promote the construction of Hong Kong CNH Centre and the internationalization of RMB

6.1   Hong Kong gold market properties

The Hong Kong gold market has had a history of over 90 years and mainly consists of the Chinese Gold & Silver Exchange, local London gold market and gold futures market. Establishment of the Chinese Gold & Silver Exchange in 1918 marked the formation of the Hong Kong gold market. In 1974, the Hong Kong government abolished the Gold Export and Import Prohibiting Ordinance, and permitted free input and output of gold. From then on, the Hong Kong gold market has developed rapidly.

The Hong Kong gold market fills the time vacuum between the closing of the New York market, and the opening of the London market, and it is convenient for international investors to continue their business in Hong Kong, be it for hedging or arbitrage. Then it links up Asia, Europe and America and ensures that the international gold business can be conducted 24 hours, constantly, in Europe, America and Asia. Its excellent geographical advantages attract European gold dealers. Five major gold dealers in London and three Swiss banks have set up branches in Hong Kong one after another. They bring part of gold trading settled in London to Hong Kong and gradually form an intangible "Local London Gold Market" with the "local London system", and make Hong Kong one of the major gold markets of the world.

6.1.1   Development process and composition of Hong Kong gold market

Compared with other major gold markets of the world, the Hong Kong gold market is an Asian market, which not only has special gold trading places, but also offers gold in the form of futures as well as spot. As early as 1904, a market-type gold and silver currency exchange had emerged in Hong Kong. Its practitioners organized "Gold and Silver Industry" (the predecessor of Chinese Gold & Silver Exchange) in 1910, and drew up simple industry regulations at that time. In 1918, Gold and Silver Industry registered with the government and officially renamed itself as "Chinese Gold & Silver Exchange".

6.1.1.1   Chinese Gold & Silver Exchange (CGSE) has characteristics of both floor trading and over-the-counter (OTC) transactions

The CGSE has a history of over 100 years, and has developed into the trading place accepted by traders, gold dealers, investors and speculators in Asia. It has expanded its organization and scale constantly and is now the only gold exchange that trades spot gold in Hong Kong.

The functions of CGSE include providing the premises, equipment and relevant services for trade of noble metals like gold and silver; drawing up and implementing business rules for gold exchange and standardizing transactions; supervising the process, clearing and settlement of trades and enforcing trading rules, ensuring the fulfillment of trading contracts; drawing up and carrying out a risk management system, and controlling the market risk; developing an official price and publishing market information; supervising and controlling trading business of members, and investigating and punishing those who violate the regulations of CGSE.

CGSE is different from the international gold market. It is a market where Chinese gold dealers are in a dominant position. It trades 99 percent gold contracts and gold bar contracts in the form of open outcry, and trades London gold and silver contracts in the electronic form, and provides the service of "trading code". Its trading unit is not the internationally accepted troy ounce but "simaliang (司馬両)".

6.1.1.2   The OTC market of "local London gold" in Hong Kong

The local London gold market in Hong Kong is the natural extension of the gold market in London. Since Hong Kong abolished restrictions on gold import and export in 1974, gold trading in CGSE has become very active, which attracted Hong Kong gold dealers' and prompted them to set up offices in Hong Kong one by one. Besides, in order to join international gold trading, many gold trading brokers in Hong Kong have agreements with gold dealers in London, which allows them to quote prices in ounce and USD as the unit in Hong Kong and receive the settlement in London.

This has resulted in the formation of the local London gold market in Hong Kong, whose trading system copies that of London gold market. It trades local London gold as international spot gold and has developed the local London gold market into the agent market of the London gold market, which has been critical to the internationalization of the Hong Kong gold market. The local London gold market in Hong Kong is an open market. At present, most of the traders are banks, investment companies and local gold dealers. They buy London gold to hedge their sales of futures. The local London gold market also accepts London gold trades transacted elsewhere in the Asia–Pacific region and two other centres that are beyond their working hours.

6.1.1.3   Futures market of gold in Hong Kong

The gold futures market is a submarket that forms a part of the metals market of the Hong Kong Stock Exchanges and Clearing Limited (HKEx), and one must entrust trade to the members of HKEx or functions as their agents to join gold trading. HKEx is a formal floor futures trading market, and like the gold futures trading market in New York and Chicago, it can effectively remedy the shortage of liquidity in the market caused by spot trading in gold and silver. The Hong Kong gold futures market has characteristics of leverage effects, low participation costs, high efficiency and transparency, strong competitiveness, low counterparty risk, formal trading mode, sound settlement system, some obligatory cash deposit by both buyers and sellers of futures and so on. Besides the above three major components, the Hong Kong gold market also includes gold coins market, gold passbook market and trading market of physical gold like ornaments.

6.1.2   Characteristics of the Hong Kong gold market

Three parts, namely, the CGSE which has developed from the traditional gold market, the local London gold market set up by branch offices of gold dealers in London and Zurich and the Hong Kong gold futures market in HKEx together form the unique Hong Kong gold market whose characteristics are as follows:

6.1.2.1   The free foreign exchange system, free trading and access across borders are allowed in Hong Kong

The Hong Kong Special Administrative Region (HKSAR) doesn't have any foreign exchange controls, allows free convertibility of HKD, and has open markets of foreign exchange, gold, securities and futures.

Hong Kong has a completely open market economy. The two stimulators of its development are no foreign exchange controls and favorable geographical location and time zone. The Hong Kong foreign exchange market is the financial market with maximum volume from Hong Kong itself. The openness of the Hong Kong gold market is largely dependent on the openness of the Hong Kong foreign exchange market which allows free convertibility of different currencies and allows prices of other gold markets to be quoted and cross-market trading. The characteristics of greater freedom and flexibility reduce the side effects of policy interventions and constraints to the lowest.

6.1.2.2   Floor trading and OTC transactions coexist, and trading in the gold market is active

Generally speaking, depending upon whether there is a tangible place or not, the gold markets of the world can be divided into floor trading markets and OTC transactions markets. For example, the London gold market is a typical OTC transactions market. And the relevant trading of gold products can be divided into standardized varieties of floor trading and non-standardized varieties of OTC transactions.

But strictly speaking, sometimes there is no clear differentiation between floor trading and OTC transactions in the international gold market. Taking the CGSE as an example, though it is usually regarded as an OTC market, there are also several characteristics of floor trading. So in fact, it can be considered a gold trading place whose trading characteristics are between OTC and floor trading. Thus, there are both floor trading and OTC transactions existing in the Hong Kong gold market, mutually separate but complementary to each other. The Hong Kong gold market is a typical new gold market system in which tangible market and intangible market coexist and futures market and spot market not only coexist but are also separate from each other. Its gold trading radiates to every corner of the world.

6.1.2.3   *Strict industry self-discipline is combined with proper government supervision*

CGSE is an independent legal entity. The congress of members is its' highest authority, and the board of directors and supervisors draws policies, conducts supervision and administration and has a set of sound industry regulations. As a trading place of gold futures, HKEx is supervised according to the Securities and Futures Ordinance by the Securities and Futures Commission of Hong Kong (SFC). The Hong Kong government implements a positive non-intervention policy on its business development and gives full freedom to the development of the gold market.

6.2   RMB Kilobar gold

The RMB Kilobar Gold Contract is a legally binding gold contract with its price fixed on board for physical settlement at a prescribed time in the morning session on every business day.

Each RMB Kilobar Gold Contract contains one lot of spot gold. The value of the RMB Kilobar Gold Contract is equal to the contracted price multiplied by the number of contracts.

RMB Kilobar Gold Contract is traded under the Electronic Trading System, which is established and operated by the CGSE. Software system vendors are subject to change at appropriate times.

At the close of each trading day, the settlement system of the CGSE calculates the book value of RMBKilobar Gold profit or loss (mark-to-market), and credits the profit and debits the loss to the accounts of contract buyers/sellers or withdraws from these accounts the amounts which are the differences between contract prices and daily settlement prices multiplied by number of contracts.

A system of liquidity providers is established to ensure the maintenance of basic liquidity in the trading of RMBKilobar Gold. Under the CGSE's trading procedures, liquidity providers are responsible for offering continuous quotes of bid prices/ask prices for a minimum number of contracts.

RMB-denominated physical gold products can provide a good channel for orderly backflow of Hong Kong RMB.

First, different from the inflow of RMB and foreign currency that needs to differentiate between current account and capital account strictly, inflow of gold is currently not restricted by any policy and gold flowing in can flow out easily as long as it is put on record. China's central bank is also adjusting the reserve policy that is over-dependent on the USD and expects the inflow of gold actively. This is determined by the specific attributes of gold. Gold will never become "hot money" in the realistic sense, its currency is not as convenient as other forms of capital (cost of transportation and conservation), and the backflow of gold will not affect the stability of RMB, so there is no need for the local central bank to hedge against it and it is manageable and controllable. Besides, because of the stability of gold value, its attraction to investors will never be affected. The specific attributes of gold make it possible that specific RMB gold products that have been certified can be accepted by the Mainland financial authority as an important channel for RMB backflow.

Second, with the acceptability of RMB-denominated gold products increasing in the world, "extracorporeal circulation" of CNH can be advanced. "Backflow System" is the bottleneck that the development of offshore market needs to break through, and the "extracorporeal circulation" system is another important aspect of protecting the long term health of the offshore market. The rapid development of the RMB gold market means that both market makers and investors have a considerable amount of RMB to get involved in transactions. This part of RMB will be preserved in Hong Kong or be traded throughout the world and have no need to return to Mainland China.

Next, RMB gold products will enhance the diversity of the RMB assets market in Hong Kong and provide a kind of investment and hedging tool. Assuming the volume of deposits of Hong Kong RMB develops rapidly, availability of investment products is rather limited and the Hong Kong gold market which is mature and active can provide good investment opportunity for a part of RMB assets.

In addition, RMB-denominated gold products will contribute to the formation of marketization indicators, and provide reference for reform of the Mainland's system of interest rate and exchange rate. Now under the condition that the formation mechanism of interest rate and exchange rate of CNH market is unsound and is affected by many factors, because of the specific attributes of gold, the gold market can sensitively reflect the fluctuation of interest rate as well as exchange rate and maintain stability at the same time. As one of the four major gold markets of the world, the mature Hong Kong gold market can be an important reference to relevant indicators. This is also a unique advantage of Hong Kong.

6.3   Gold trading center in Qian Hai (Shen Zheng)

China's gold management system in the present stage can be further improved. First, the examination and approval system for import and export of gold and gold products is still in force. This is far from the goal of relaxation of gold import and export controls and realization of complete marketization and liberalization of gold imports and exports. Second, China's gold market represented by Shanghai Gold Exchange is still a relatively semi-closed market since gold is only imported, not exported. Exports need to go through strict examination and approval. The onshore CNY funds are used for trade. The trading currency is not freely convertible.

The Chinese Gold & Silver Exchange Society enjoys considerable advantages because of the Gold Trading Center in Qianhai. First, "the moving of Western gold to the East" has become an irreversible trend in the international gold market. Asia–Pacific countries are in urgent need of a gold trade center in the region to meet the robust demand. At present, much of the world imports of gold are from areas other than the producing areas and the gold so traded is known as transit gold. For Hong Kong/Qianhai, after a trading transaction, gold can either go to China or be transited to other countries. With the establishment of this free trade zone, a rather convenient window for trade, international transactions in gold have become more convenient. This will help Hong Kong/Qianhai to become a major gold transit center in the Asia–Pacific region and the world. Second, after nearly 20 years of development, the gold and jewelry industry has preliminarily established its position in the international market. At present, Shenzhen houses more than 70 percent of gold and jewelry enterprises in China, which process more than 1,000 tonnes of gold each year. It has an important position and influence in the world. As a world-famous gold and jewelry processing center, Shenzhen is in urgent need of a production factor market that can match with it. Operating in the international market-driven mode, the Qianhai Gold Trading Center serves the gold and jewelry industry in Shenzhen, as well as the Pearl River Delta and South China, which can significantly save transaction costs, shorten the transaction time and improve transaction efficiency. Third, the Chinese Gold & Silver Exchange Society has a high degree of marketization and internationalization, a lot of experience and mature and standardized operations. Fourth, Qianhai has a geographically advantageous location and infrastructures in Hong Kong as well as Shenzhen are of global standards. The modern port, warehousing, logistics, communications, finance and other service industries in this area are in leading positions in China and even the world. The normal, healthy and orderly operation of a gold trading center can be ensured.

Therefore, the Qianhai Gold Trading Center should operate as a trade service center integrating transactions, delivery, settlement and storage, taking advantage of gold import and export policies and added-value tax rules in China. It is an improved version of, and supplements, the Shanghai Gold Exchange. Two trading systems complementing each other, competing with each other, having obvious differences and their own unique features are developing together. The Shanghai Gold Exchange is inside the special economic zone in China, and the Qianhai Gold Trading Center is outside the special economic zone in China. Seen from the current situation, Hong Kong/Qianhai has the ability to become a transit center. For demand from China and other Asia–Pacific countries, transit in Qianhai has more advantages in terms of both time and cost, so the corresponding international investors hope to come to Qianhai for gold transit. More international gold production enterprises can be expected to deliver gold to the warehouses of the Qianhai Free Trade Zone before transit in coming years. To build a gold transit center in the Asia–Pacific region, a gold warehouse has already been set up independently in the free trade zone. It can provide a range of supporting services, including delivery, warehousing, clearing and logistics. If its operational stocks reach a certain level, the Qianhai Spot Gold and Silver Trading Platform will be able to fully meet the requirements of international delivery, and is expected to become Asia's largest precious metals storage warehouse.

Seen from experience, the agglomeration effect of the CNY gold market is obvious. More international investments and offshore CNY funds will participate in domestic and foreign financial market transactions, resulting in greater integration of offshore and onshore institutions, offshore and onshore CNY and increased rate of fund flows. The linkages between gold, interest rate and exchange rate market shall be strengthened, which shall vigorously promote the role of the offshore financial markets in efficient resource allocation and the construction of the international financial center.

6.4 The arbitrage model of gold on offshore and onshore market

6.4.1 Background

Since the CNY settlement in cross-border trade was launched in July 2009, the exchange rate and interest rate of offshore CNY have been formed in CNH and they are freely floating, influenced by market and the exchange rate and interest rate on the Mainland. Exchange rate and interest rate on the Mainland are now different, providing the basis for cross-border arbitrage.

From the view point of development history of CNH, it is normal for the two-way volatility of exchange rate and there have been significant differences during some stages. From the second half of 2010 to the first half of 2011, the spot rate of Hong Kong offshore CNY was significantly higher than onshore CNY, especially from September to October in 2010; the highest offshore CNY exchange rate is 1700 basis points (BPs) higher than that of onshore CNY.

6.4.1.1 Arbitrage of exchange rate differential

During the early days of CNY development in Hong Kong, CNY business developed slowly. After the People's Bank of China loosened monetary policy in August 2010 and signed with the Bank of China (Hong Kong) a new settlement agreement, CNY deposits in Hong Kong enjoyed an explosive growth. The main reason is that CNY appreciation expectation at that time was powerful; the exchange rate of Hong Kong offshore CNY had been higher than the onshore market in Mainland China for a long time. Therefore, when importers from the Mainland make payments to foreign countries they transfer the foreign exchange purchasing spontaneously to Hong Kong to reduce costs by making use of CNH. Overseas export enterprises directly sell CNY on the offshore market after receiving it for more foreign exchange funds.

A CNY gold contract is denominated in CNY, while a USD gold contract is denominated in USD. There is an implied CNY exchange rate between the two (hereinafter referred to as implied CNY exchange rate). The implied exchange rate fluctuates above and below the offshore CNY exchange rate. Because the implied CNY exchange rate of gold is calculated with the price of gold, any factor of gold price will affect the implied CNY exchange rate of gold. Ideally, according to the law of one price, after adjustment the domestic gold price and overseas gold price should be equal. On this basis, the theoretical price of gold in CNY can be calculated with the following formula:

The theoretical price of gold in CNY = price of gold in USD X offshore CNY exchange rate + shipping cost + insurance premium + tax + refining cost

Here, refining cost refers to the conversion cost of different gold standards (99 and 9999), and gold can meet delivery standard only after refining; tax mainly refers to tariff, and gold tariff in all major countries is 0; and shipping cost and insurance premium are the transportation costs of gold. Normally, the total proportion of transportation cost, insurance premium and refining cost is not more than 0.125 percent. The effect is very small and can be ignored.

By holding different gold portfolios, a variety of CNY implied exchange rates can be formed. When the implied exchange rate has a large deviation with RMB offshore market price, interest arbitrage can be carried out. In addition to the spot market of CNY, forwards, options and other actual interest arbitrage varieties can be considered. Since the exchange rate reform in August, the CNY market fluctuations have increased and interest rate arbitrage opportunities have also significantly increased. By grasping the day trading time point and selecting highly leveraged products, the level of investment can be significantly improved.

When CNY is not completely freely convertible, the domestic commodity price represented by gold cannot reflect CNY exchange rate market information in time. It becomes more obvious after the foreign exchange trading center closes and the market has interest arbitrage opportunities. Enterprises or investment institutions that need to hedge against the risk of the CNY exchange rate can obtain additional revenue by seizing the above interest arbitrage opportunities. At the same time, interest arbitrage actually strengthens the linkage between the gold market and CNY exchange rate market both at home and abroad, improves the offer efficiency of relevant markets, and helps CNY to accelerate the process of internationalization and free conversion.

This kind of arbitrage model requires a real background of import and export trade and is thus restricted by such transaction costs as transportation and insurance costs and the scale in the initial stages was not large. However, Mainland China found that the product transaction cost of this industry is low, and they can evade the limitations of customs through the free trade zone. Therefore, by creating the background of import and export trade, arbitrage circulation accelerated and large arbitrage profits can be generated. First arbitrage enterprises use their own funds in the Mainland or borrow $1 million and then exchange it for CNY 6.2 million at the exchange rate of 6.20 on the onshore market. Then in the name of importing raw materials they import goods with low logistics cost, such as gold from the related company in Hong Kong, and pay in CNY to foreign enterprises. The 6.2 million of CNY remitted to Hong Kong by arbitrage enterprises become an offshore RMB deposit. Next, arbitrage enterprises affiliated with Hong Kong's related company exchange CNY for USD at the offshore exchange rate of 6.15 to pay RMB for dollars (CNY exchange rate on offshore market is higher than that of onshore) and they can get USD 1.00813 million. Finally, arbitrage enterprises in Mainland China export the processed gold to the related company in Hong Kong and settle accounts in USD, gaining a payment for goods of USD 1.00813 million. The final account returned to the Mainland is USD 1.00813 million. The whole gold arbitrage of imported raw material and exported manufactured goods is completed with a profit of USD 8130.

As the transit trade centre between inland and overseas, after the foreign exchange purchase in Hong Kong CNY is remitted abroad, and thus CNY is gradually deposited in Hong Kong. Meanwhile, Hong Kong and Mainland banks have developed a series of related products, such as the non-landing remittance for CNY, which to some extent further accelerates the convenience and speed of arbitrage.

6.4.2   The influence of arbitrage

6.4.2.1   Positive influence

ARBITRAGE INCREASES THE SCALE OF CNY DEPOSITS IN OFFSHORE MARKET

When the exchange rate of CNH is higher than the onshore market, arbitrage objectively promotes it at home and abroad, especially to those who use CNH, instead of CNY, as the method of settlement which has promoted the business scale of CNY settlement in cross-border trade and accelerated payments by domestic enterprises to overseas enterprises in CNY. This increases CNY stock rapidly in the Hong Kong market in the short term and expands CNH.

At the initial stages, the CNY business in Hong Kong developed slowly, and after the new settlement agreement took effect CNY deposits in Hong Kong enjoyed an explosive growth. The main reason is that CNY appreciation expectation at that time was powerful; the exchange rate of Hong Kong offshore CNY had been higher than the onshore market in Mainland China for a long time, which kept a definite gap.

Enterprises in Mainland China transfer the foreign exchange purchasing spontaneously to Hong Kong to reduce costs by making use of CNH. Or parts of overseas export enterprises directly sell CNY on the offshore market after receiving it for more foreign exchange funds.

If the exchange rate on the onshore market is higher than the rate for CNH, arbitrage promotes the currency at home and abroad, especially when enterprises in the Mainland choose CNY settlement and sell foreign exchange to buy CNH in Hong Kong. Therefore it in fact promotes the business scale of CNY settlement in cross-border trade, accelerates the purchasing of CNY by foreign enterprises on the overseas offshore market and the companies in the Mainland pay CNY as the currency of settlement, resulting in a decline of the capital stock of CNY in the Hong Kong market but also activating the exchange rate of CNH.

Therefore, the two-way arbitrage has promoted the amount of CNY settlement in CNH and improved the trade volume of foreign exchange. But the influence of CNY deposit balance is just the opposite. As a whole, there is still a lot of room for appreciation of CNY, so the CNY deposit scale in CNH has always been at the expansion stage.

ARBITRAGE INCREASES THE FINANCING SCALE OF OFFSHORE MARKET

The original arbitrage of the Mainland companies actually promotes transference of financing behavior to the related company of Hong Kong, which obtains the financing from the Hong Kong bank, which significantly improves the trade finance growth in CNH. On the other hand, in order to increase income, arbitrage enterprises use this model repeatedly and use the leverage to enlarge the volume of financing, which has a huge impact on the trade financing balance of Hong Kong banks, leading to the Chinese government's powerful supervision in the Mainland. Moreover the Hong Kong Monetary Authority inspects the related trade financing activity in Mainland China and Hong Kong banks at the same time. Because this kind of financing is generally in either USD or HKD, and it uses letter of credit issued by the Mainland as collateral, so the growth rate of this kind of financing can be examined by checking the volume of claim to mainland banks in the Hong Kong banking system. From the analysis of the creditors' rights of banking system in Hong Kong to the Mainland bank published by Hong Kong's financial management on its website, it can be seen that during the period 2004 to 2009, the creditors' rights of Hong Kong banks to Mainland banks almost stayed unchanged with a total of HKD $500 billion. Since 2010, the creditors' rights rose rapidly and in June 2013 it reached HKD 2.1 trillion. The main reason may be related to the arbitrage that expanded financing in Hong Kong.

6.4.2.2 The negative impact

There are three aspects to the negative impacts of arbitrage. The first is that hot money flows into Mainland China, influencing money supply and the effect of macro-control. The second is that with the expansion of the arbitrage scale, it has had a large impact on China's import and export statistical data in early 2013, leading to the Chinese government's powerful supervision on arbitrage. The third is that offshore markets affect the CNY exchange rate of onshore markets.

6.4.3   Policy and suggestion

Although the most controversial cross-border arbitrage that influences further development of offshore market in Hong Kong has little effect on money supply on the Mainland and the exchange rate, it helps promote development of CNH and internationalization of CNY. It is recommended that the Chinese government should not stop the support of CNH just because of arbitrage but continue to steadily push forward the internationalization of CNY and the construction of CNH.

6.4.3.1   Rational utilization of market arbitrage to continue the expansion the CNH scale

There were several stages when development slowed down during the process of CNH. The moment of the difficulty mainly lies in the Chinese government's policy that the government expand the scope of CNY settlement in cross-border trade and loosen foreign exchange management and capital controls. Therefore, in the early days of the CNY offshore market it was more dependent on government policy. Market-oriented policy can promote the development of the offshore market quickly.

These policies, to some extent, led to interest rate differentials between onshore and offshore markets and proliferation of exchange rate arbitrage. But supervision of arbitrage by the Chinese government has not fully returned to the degree where it manages international payments of CNY according to the foreign exchange management mechanisms such as the dollar.

At present, arbitrage can be regarded as a kind of cost to promote internationalization of the CNY. Therefore, it is suggested that in the early development stage of CNH, in theory there be no payment difficulties for CNY-denominated debt. CNY policy should be further relaxed and further expansion of CNH scale needs to be promoted.

6.4.3.2   Further expand investment in the offshore market in Hong Kong

For development of the investment market in Hong Kong, such as bonds, stocks and funds, market participants from the Mainland are important. As a free economy, for issue of bonds, stocks and funds, Hong Kong adopts a registration system, that is, investors are at their own risk, which is significantly different from the examination and approval system in Mainland China. Based on the market-oriented principle with investors bearing the risk, further loosening of limitations on Mainland enterprises to raise financing in Hong Kong is suggested. Rapid expansion of the investment market in Hong Kong should be encouraged. In the process, the experience of Hong Kong can be drawn upon to promote the bond market in Mainland China and for reform of stock market for initial public offerings (IPOs).

Bibliography

Adrangi, Bahram, Chatrath, Arjun, and Christie-David, Rohan, Price discovery in strate-gically-linked markets: The case of the gold-silver spread, *Applied Financial Economics*, Vol. 10, Issue 3, 2000, 227–234.

Aggarwal, R. and Soenen, L.A., The nature and efficiency of the gold market, *Journal of Portfolio Management*, Vol. 14, Issue 3, 1988, 18–21.

Akgiray, V., Booth, G., Hatem, J., and Mustafa, C., Conditional dependence in precious metal prices, *Financial Review*, Vol. 26, Issue 3, 1991, 367–386.

Baffes, John, Oil spills on other commodities, *Resources Policy*, Vol. 32, Issue 3, September 2007, 126–134.

Baker, S.A., and Van-Tassel, R.C., Forecasting the price of gold: A fundamentalist approach, *Atlantic Economics Journal*, Vol. 4, Issue 13, 1985, 43–52.

Barkoulas, Jolln T., Hu, Aidong, and Santos, Michael R., The link between commodity prices and commodity-linked equity values during a geopolitical event, *Academy of Accounting and Financial Studies Journal*, Vol. 12, Issue 2, 2008, 1–26.

Baur, Dirk G. and McDermott, Thomas K., Is gold a safe haven? International evidence, *Journal of Banking & Finance*, Vol. 34, Issue 8, August 2010, 1886–1898.

Bertus, M. and Stanhouse, B., Rational speculative bubbles in the gold futures market: An application of dynamic factor analysis, *The Journal of Futures Markets*, Vol. 21, Issue 1, 2001, 79–108.

Bhar, R. and Hamori, S., Information flow between price change and trading volume in gold futures contracts, *International Journal of Business and Economics*, Vol. 3, Issue 1, 2004, 45–56.

Blose, Laurence E., Gold price risk and the returns on gold mutual funds, *Journal of Economics and Business*, Vol. 48, Issue 5, December 1996, 499–513.

Blose, Laurence E., How surprise changes in inflationary expectation affect the gold price, *International Advances in Economic Research*, Vol. 6, Issue 2, 2000, 369.

Blose, Laurence E., Gold prices, cost of carry, and expected inflation, *Journal of Economics and Business*, Vol. 62, 2010, 35–47.

Burton, Paul, Primary gold supply – countries, companies, consolidation and costs, *Applied Earth Science*, Vol. 114, Issue 2, 2005, 108–114.

Cai, J., Cheung, Y.L., and Wong, M.C.S., What moves the gold market?, *Journal of Futures Markets*, Vol. 21, Issue 3, March 2001, 257–278.

Capie, Forrest, Mills, Terence C., and Wood, Geoffrey, Gold as a Hedge against the US Dollar, *Journal of International Financial Markets, Institutions and Money*, Vol. 15, Issue 4, 2005, 343–352.

Cheung, Y. and Lai, K.S., Do gold market returns have long memory?, *The Financial Review*, Vol. 28, Issue 2, 1993, 181–202.

Chow, Ying-Foon, Arbitrage, risk premium, and cointegration tests of the efficiency of futures markets, *Journal of Business Finance & Accounting*, Vol. 28, Issues 5–6, 2001, 693–713.

Christie–David, R., Chaudhry, M., and Koch, T.W., Do macroeconomics news releases affect gold and silver prices?, *Journal of Economics and Business*, Vol. 52, Issue 5, 2000, 405–421.

Ciner, C., On the long run relationship between gold and silver prices a note, *Global Finance Journal*, Vol. 12, Issue 2, Autumn–Winter 2001, 299–303.

Cohen, Benjamin J., The seigniorage gain of an international currency: an empirical test. *Quarterly Journal of Economics,* Vol. 85, Issue 3, 1971, 494–507.

Dempster, Natalie, *Investment Research Manager, Investing in Gold: The Strategic Case*, London: The World Gold Council, 2008.

Dhillon, Upinder S., Lassera, Dennis J., and Watanabe, Taiji, Volatility, information, and double versus walrasian auction pricing in US and Japanese futures markets, *Journal of Banking & Finance*, Vol. 21, Issue 7, July 1997, 1045–1061.

Dooley, M.P., Isard, P., and Taylor, M.P., Exchange rates, country specific shocks and gold, *Applied Financial Economics*, Vol. 5, Issue 3, 1995, 121–129.

Edel, T., Seasonality, risk and return in daily COMEX gold and silver data 1982–2002, *Applied Financial Economics*, Vol. 16, Issue 4, February 2006, 319–333.

Edel, Tully and Brian, Lucey M., A power GARCH examination of the gold market, *Research in International Business and Finance*, Vol. 21, Issue 2, June 2007, 316–325.

Einzig, P., The forward price of gold, *The Economic Journal*, Vol. 48, Issue 192, 1938, 748–751.

Engle, R.F. and Granger, Clive W.J., Co-integration and error correction: Representation, estimation, and testing, *Econometrica: Journal of the Econometric Society*, Vol. 55, Issue 2, 1987, 251–276.

Fama, E.F. and French, K.R., Business cycles and the behavior of metals prices, *The Journal of Finance*, Vol. 43, Issue 5, December 1988, 1075–1093.

Fortune, J.N., The inflation rate of the price of gold, expected prices and interest rates, *Journal of Macroeconomics*, Vol. 9, Issue 1, 1987, 71–82.

Garber, P., What currently drives CNH market equilibrium, The Council on Foreign Relations and China Development, Research Foundation Workshop on the Internationalization of the Renminbi, Beijing, 2011–10–31.

Harmston, Stephen, *Gold as a Store of Value*. London: The World Gold Council, 1998.

Hartmann, P., *Currency Competition and Foreign Exchange Market: The Dollar, the Yen, and the Euro*, Cambridge: Cambridge University Press, 24, 1998.

Hillier, D., Draper, P., and Faff, R., Do precious metals shine: An investment perspective, *Financial Analysts Journal*, Vol. 62, Issue 2, March–April 2006, 98–106.

Ivanova, K. and Ausloos, M., Low q-moment multifractal analysis of gold price, Dow Jones industrial average and BGL-USD exchange rate, *The European Physical Journal B*, Vol. 8, Issue 4, 1999, 665–669.

Iwami, T. The Internalization of the Yen, and Key Currency Questions, *Japan in the International Financial System*. London: Palgrave Macmillan, 1995, 118–150.

Jastram, R.W., *The Golden Constant: The English and American Experience, 1560–1976*, New York: Wiley, 1977.

Jonathan, A.B. and Brian, M.L., Volatility in the gold futures market. Institute for International Integration Studies, 2007, Dublin, Vol. 6: 101–113.

Kennedy, Steven C., *A Historical Review: American Gold Market*. London: The World Gold Council, 2002.

Koutsoyiannis, A., A short-run pricing model of a speculative asset tested with data from the gold bullion market, *Applied Economics*, Vol. 15, Issue 5, October 1983, 563–582.

Lauterbach, B. and Monroe, M., Evidence on the effect of information and noise trading on intraday gold futures returns, *The Journal of Futures Markets*, Vol. 9, Issue 4, 1989, 297–305.

Lawrence, Colin, *Why Is Gold Different from Other Assets? An Empirical Investigation*. London: The World Gold Council, March 2003.

Levin, Eric J. and Wright, Robert E., *Short-Run and Long-Run Determinants of the Price of Gold*. London: The World Gold Council, June 2006.

Liu, S.M. and Chou, C.H., Parities and spread trading in gold and silver markets: A fractional cointegration analysis, *Applied Financial Economies*, Vol. 13, 2003, 899–911.

Mahdavi, S.S., and Zhou, Su, Gold and commodity prices as leading indicators of inflation: Tests of long-run relationship and predictive performance, *Journal of Economics and Business*, Vol. 49, Issue 5, September–October 1997, 475–489.

Malliaris, A.G. and Malliaris, Mary, Time series and neural networks comparison on gold, oil and the Euro, Proceedings of International Joint Conference on Neural Networks, June 14–19, 2009, Atlanta, Georgia, USA: 1961–1967.

Marcuzzo, M.C. and Rosselli, A., Profitability in the international gold market in the early history of the gold standard, *Economica, New Series*, Vol. 54, Issue 215, August 1987, 367–380.

Maziad, S. and Kang, J.S., RMB Internationalization: Onshore/Offshore Links. IMF Working Paper, WP/12/133, May 2012.

McKinnon, Ronald, The rules of the game: International money in historical perspective, *Journal of Economic Literature*, Vol. 1, Issue 31, 1993, 1–44.

Melvin, M. and Sultan, J., South Africa political unrest, oil prices, and the time varying risk premium in the gold futures market, *Journal of Futures Markets*, Vol. 10, Issue 2, April 1990, 103–112.

Michaud, Richard, Michaud, Robert, and Pulvermacher, Katharine, *Gold as a Strategic Asset*, London: The World Gold Council, 2006.

Mills, Terence C., Statistical analysis of daily gold price data, *Physica A: Statistical Mechanics and Its Applications*, Vol. 338, Issues 3–4, 15 July 2004, 559–566.

Monroe, M.A. and Cohn, R.A., The relative efficiency of the gold and treasury bill futures markets, *The Journal of Futures Markets*, Vol. 6, Issue 3, 1986, 477–493.

Mundell, Robert, Does Africa need a common currency? *The Third African Development Forum: Defining Priorities for Regional Integration*, 2002, 45–57.

Murase, T., Hong Kong Renminbi Offshore Market and Risks to Chinese Economy. *Institute for International, Monetary Affairs*, Newsletter No. 40, 2010. http://www.iima.or.jp/Docs/newsletter/2010/NLNo_40_e.pdf

Pulvermacher, Katharine, *Analysis of Long-Run Correlation of Returns on Gold and Equity*. Center for Public Policy Study, London: The World Gold Council, 2003.

Quadrio-Curzio, Alberto, *The Gold Problem: Economic Perspectives*, 1st Edition, Oxford, USA: Oxford University Press, August 25, 1983.

Ranson, David and Wainwright, H.C., *Inflation Protection: Why Gold Works Better Than "Linkers"*. London: The World Gold Council, November 2005.

Salant, S.W. and Henderson, D.W., Market anticipations of government policies and the price of gold, *Journal of Political Economy*, Vol. 86, Issue 4, August 1978, 627–648.

Sato, Kiyotaka, The international use of the Japanese Yen: The case of Japan's trade with East Asia, *The World Economy*, Vol. 22, Issue 4, June 1999, 547–584.

Sherman, E.J., A gold pricing model, *Journal of Portfolio Management*, Vol. 9, 1983, 68–70.

Shigui, Tao, New path to promote RMB's international acceptability, *Finance and Economics*, Vol. 6, 2009, 1–9.

Sjaastad, Larry A. and Scacciavillani, Fabio, The price of gold and the exchange rate, *Journal of International Money and Finance*, Vol. 15, Issue 6, December 1996, 879–897.

Sjaastad, Larry A., The price of gold and the exchange rates: Once again, *Resources Policy*, Vol. 33, Issue 2, June 2008, 118–124.

Smith, Graham, *The Price of Gold and Stock Price Indices for the United States*. London: The World Gold Council, 2001.

Smith, Graham, *London Gold Prices and Stock Price Indices in Europe and Japan*. London: The World Gold Council, 2002.

Solt, Michael E. and Swanson, Paul J., On the efficiency of the markets for gold and silver, *The Journal of Business*, Vol. 54, Issue 3, July 1981, 453–478.

Spyrou, S., Unobservable information and behavioral patterns in futures markets: The case for brent crude oil, gold and robusta coffee contracts, *Derivatives Use, Trading & Regulation*, Vol. 12, Issues 1–2, 2006, 48–55.

Tandon, K. and Urich, T., International market response to announcements of US macroeconomic data, *Journal of International Money and Finance*, Vol. 6, Issue 1, March 1987, 71–83.

Tavlas, G.S., The international use of the US Dollar: An optimum currency area perspective, *The World Economy*, Vol. 20, Issue 6, September 1997, 709–747.

Taylor, Nicholas J., Precious metals and inflation, *Applied Financial Economics*, Vol. 8, Issue 2, 1998, 201–210.

Triffin, Robert, The size of the nation and its vulnerability to economic nationalism. In Robinson, E.A.G. (Ed.), *Economic Consequences of the Size of Nations*. London: Palgrave Macmillan, 1960, 247–264.

Tschoegl, A.E., Efficiency in the gold market, *Journal of Banking & Finance*, Vol. 4, Issue 4, December 1980, 371–379.

Urich, Thomas J., Modes of fluctuation in metal futures prices, *Journal of Futures Markets*, Vol. 20, Issue 3, March 2000, 219–241.

Wang, Chengbiao, Chen, Yanhui, and Li, Lihong, The forecast of gold price based on the GM(1,1) and Markov chain, Proceedings of 2007 IEEE International Conference on Grey Systems and Intelligent Services, 2007, Nanjing, China, Vols. 1–2: 739–743.

Wang, Q., The speech on the RMB internationalization roundtable workshop in Hong Kong. Sponsored by Hong Kong Monetary Authority, 2011-5-23.

Xu, X.E. and Fung, H.G., Cross-market linkages between U.S. and Japanese precious metals, futures trading, *Journal of International Financial Markets, Institutions and Money*, Vol. 15, 2005, 107–124.

Yongding, Yu, The current RMB exchange rate volatility and RMB internationalization, *International Economic Review*, Vol. 1, 2012, 18–26.

Zhang, Jinliang, Tang, Mingming, and Tao, Mingxin, Effects simulation of international gold prices on crude oil prices based on WBNNK model, 2009 ISECS International Colloquium on Computing, Communication, Control, and Management, Sanya, China, 2009: 459–464.

Index

Page numbers in italic format indicate figures and tables.